INNOVATIONS IN ANTIVIRAL DEVELOPMENT AND THE DETECTION OF VIRUS INFECTIONS

ADVANCES IN EXPERIMENTAL MEDICINE AND BIOLOGY

Recent Volumes in this Series

Volume 309A
PURINE AND PYRIMIDINE METABOLISM IN MAN VII, Part A: Chemotherapy, ATP Depletion, and Gout
Edited by R. Angus Harkness, Gertrude B. Elion, and Nepomuk Zöllner

Volume 309B
PURINE AND PYRIMIDINE METABOLISM IN MAN VII, Part B: Structural Biochemistry, Pathogenesis, and Metabolism
Edited by R. Angus Harkness, Gertrude B. Elion, and Nepomuk Zöllner

Volume 310
IMMUNOLOGY OF MILK AND THE NEONATE
Edited by Jiri Mestecky, Claudia Blair, and Pearay L. Ogra

Volume 311
EXCITATION-CONTRACTION COUPLING IN SKELETAL, CARDIAC, AND SMOOTH MUSCLE
Edited by George B. Frank, C. Paul Bianchi, and Henk E. D. J. ter Keurs

Volume 312
INNOVATIONS IN ANTIVIRAL DEVELOPMENT AND THE DETECTION OF VIRUS INFECTIONS
Edited by Timothy M. Block, Donald Jungkind, Richard Crowell, Mark Denison, and Lori R. Walsh

Volume 313
HEPARIN AND RELATED POLYSACCHARIDES
Edited by David A. Lane, Ingemar Björk, and Ulf Lindahl

Volume 314
CELL-CELL INTERACTIONS IN THE RELEASE OF INFLAMMATORY MEDIATORS: Eicosanoids, Cytokines, and Adhesion
Edited by Patrick Y-K Wong and Charles N. Serhan

Volume 315
TAURINE: Nutritional Value and Mechanisms of Action
Edited by John B. Lombardini, Stephen W. Schaffer, and Junichi Azuma

---

A Continuation Order Plan is available for this series. A continuation order will bring delivery of each new volume immediately upon publication. Volumes are billed only upon actual shipment. For further information please contact the publisher.

INNOVATIONS IN ANTIVIRAL DEVELOPMENT AND THE DETECTION OF VIRUS INFECTIONS

Edited by

Timothy M. Block and Donald Jungkind

Thomas Jefferson University Hospital
Philadelphia, Pennsylvania

Richard L. Crowell

Hahnemann University School of Medicine
Philadelphia, Pennsylvania

Mark Denison

Vanderbilt University
Nashville, Tennessee

and

Lori R. Walsh

Mercy Catholic Medical Center
Darby, Pennsylvania

SPRINGER SCIENCE+BUSINESS MEDIA, LLC

Library of Congress Cataloging-in-Publication Data

Innovations in antiviral development and the detection of virus infections / edited by Timothy M. Block ... [et al.].
p. cm. -- (Advances in experimental medicine and biology ; v. 312)
"Proceedings of the Eastern Pennsylvania Branch of the American Society for Microbiology Symposium of Innovations in Antiviral Development and the Detection of Virus Infections, held November 15-16, 1990, in Philadelphia, Pennsylvania"--T.p. verso.
Includes bibliographical references and index.
ISBN 978-1-4613-6533-4 ISBN 978-1-4615-3462-4 (eBook)
DOI 10.1007/978-1-4615-3462-4
1. Virus diseases--Molecular aspects--Congresses. 2. Virus diseases--Treatment--Congresses. 3. Virus diseases--Diagnosis--Congresses. I. Block, Timothy M. II. American Society for Microbiology. Eastern Pennsylvania Branch. III. Eastern Pennsylvania Branch of the American Society for Microbiology Symposium of Innovations in Antiviral Development and the Detection of Virus Infections (1990 : Philadelphia, Pa.) IV. Series.
[DNLM: 1. Antiviral Agents--congresses. 2. HIV Infections--prevention & control--congresses. 3. Virus Inhibitors--congresses. W1 AD559 v. 312 / QV 268.5 I58 1990]
RC114.5.I52 1992
616.9'25--dc20
DNLM/DLC
for Library of Congress 92-14364
CIP

Proceedings of the Eastern Pennsylvania Branch of the American Society for Microbiology Symposium of Innovations in Antiviral Development and the Detection of Virus Infections, held November 15-16, 1990, in Philadelphia, Pennsylvania

ISBN 978-1-4613-6533-4

Timothy Block dedicates his contributions to his wife, Joan, his parents, Joyce and Carl, as well as to Russell Schaedler, M.D. and Baruch Blumberg, M.D., Ph.D. for their inspiration.

Donald Jungkind dedicates his contributions to his patient and understanding wife, Kim, and to Earle H. Spaulding, Ph.D., who was influential in his decision to study diagnostic microbiology and virology.

SYMPOSIUM SPONSORS

AMERICAN CYANAMID COMPANY

BECTON DICKINSON MICROBIOLOGY SYSTEMS

BOEHRINGER INGELHEIM PHARMACEUTICALS, INC.

BOEHRINGER MANHEIM

BURROUGHS WELLCOME

CETUS CORPORATION

DAIICHI PHARMACEUTICAL CORPORATION

ORTHO DIAGNOSITIC SYSTEMS

E. R. SQUIBB PHARMACEUTICAL COMPANY

STERLING DRUG COMPANY

UPJOHN COMPANY

WHITAKER BIOPRODUCTS, INC.

WYETH LABORATORIES, INC.

ACKNOWLEDGEMENTS

The editors wish to thank Margie Pike for her patience and tireless efforts in the administrative preparation of these manuscripts. Her influence is evident throughout the book. We also wish to thank Mary Cohen and Elham Bshouti for their typing contributions.

FOREWORD

THE ERA OF ANTIVIRALS

Introduction

Although there are more than one hundred medically useful antibiotics and fungicides, there are only seven compounds licensed for use as antiviral agents, in the USA. Some of these (acyclovir and ganciclovir) are actually derivatives of each other, making the number of new discoveries even smaller. Moreover, most of these agents are of only limited therapeutic value and have substantial toxicity.

It has been more than 100 years ago since Pasteur studied rabies virus (2) and Rous (4) showed that a small filterable agent (not bacteria) caused disease (sarcoma) in chickens. It was nearly 100 years ago that yellow fever virus, the first recognized human pathogenic virus, was unambiguously associated with disease (3). Enteroviruses were cultured for the first time nearly 50 years ago (1). Why then has effective chemotherapy against viruses lagged behind that of other microorganisms?

Viruses are often difficult to grow and image. However, with the dynamic advances in molecular biology and increased sophistication in tissue culture, the field of virology has blossomed and resulted in improved methods for detection of virus infection. The use of viruses as models of gene regulation and replication has also resulted in a massive accumulation of information. The rapid advances in molecular biology, including the discovery of specific virus functions that differ greatly from the host and of systems in which virus gene products are expressed, has liberated the study of some viruses, such as human immunodeficiency virus (HIV) and herpes simplex, from a dependency upon tissue culture growth. The discovery of virus functions and replication strategies that differ from their host has made it possible to begin to specifically target antiviral agents and overcome the main obstacle that designers of antiviral chemotherapy have faced: to create antiviral agents that block virus replication but do not substantially interfere with host cell synthetic pathways.

These recent developments give researchers confidence that in the near future many antiviral agents will be made available for clinical use. Indeed, from a chemotherapeutic perspective, it is likely that the 1990s may be called the "decade of antivirals." The meeting that this book is based upon, as well as the proceedings themselves, are testimony to the prediction that we are reaching a "critical mass" where the possibility of rapid diagnosis of viral infection plus effective therapy will result in more widespread demand for these clinical services.

This publication differs from more conventional reports of virus chemotherapy in that the authors have been encouraged to be speculative. In addition to the presentation of data, authors were welcome to make conjecture about experiments that had not necessarily been conducted and to hypothesize about approaches that have not yet been taken. We have emphasized innovations because with only a handful of antiviral agents out there, some new approaches might be helpful.

Highlights of the Book

Some of the highlights of the proceedings are outlined here, although it is noted that not all of the contributions to this book are considered in this forward.

Antiviral strategists graphically divide the virus life cycle into discreet steps as outlined below:

Step 0	Processing of the virion in host biological fluids and access to target tissue
Step I	Attachment
Step II	Entry and uncoating
Step III	Relocation of the viral genome to appropriate cell compartments for synthetic activity
Step IV	Transcription of the viral genome and translocation of viral mRNA, inhibition of host macromolecular synthesis or transformation events
Step V	Replication
Step VI	Assembly and egress
Step VII	Immunological effects

Although these steps are somewhat artificial, in the sense that the virus life cycle is actually more of a continuum than a sequence of separately occurring events, they do help organize our thinking. Indeed, the findings described in this book can, in some regards, be categorized by the steps that are interfered with. For example, categorizing certain chapters by the step in the virus life cycle they discuss allows us to offer the following outline:

Step 0: Processing of the Virus and Access to Host Tissue

The book starts with an important lesson and reminder from Bernard Fields. Dr Fields emphasizes the limitations of studying viruses in tissue culture and provides several examples of steps in a virus' life cycle that could only have been revealed by *in vivo* analysis of pathogenesis. For example, the proteolytic processing of reovirus in the mouse gut, a potentially useful target for antiviral action, would not have been appreciated if analysis of the virus life cycle had been limited to tissue culture. Dr. Bernard Fields reports that reovirus is proteolytically cleaved in the mouse gut. This cleavage results in the "activation" of virus attachment proteins, allowing for efficient binding to the host cell. Inhibition of this processing event results in effective interference with virus replication *in vivo*. The dynamics of virus spread throughout its host are also reviewed. Remarkably, antibody against reovirus administered to mice that already harbor the virus in their central nervous system can still drastically reduce the spread of virus throughout specific neurotracts. Other full-length contributions that deal with an aspect of virus pathogenesis are by Wu et al., and Coen et al.

Step 1: Attachment

Receptors for virus attachment provide very attractive antiviral targets. In particular, Richard Colonno reviews the use of anti-rhinovirus receptor agents in the context of the knowledge that the functional receptor is immune cell adhesion molecule #1. The cellular receptor for HIV has also been targeted. An approach that targets the virion, rather than the host, is outlined by Berger et al. Briefly, for example, soluble CD4, the cellular receptor for HIV, has been derivatized and used to "neutralize" *in vitro* preparations of the virus. The potential of this agent, in the context of combination therapy, is presented. Another full-length contribution that focuses upon virus receptors is by Dutko et al., in which the possibility of using antiviral agents of known mechanisms of action to probe virus function is discussed.

Step II: Entry and Uncoating

This step offers opportunities for antiviral action, but is not considered in this report.

Steps III and IV: Transcription and Replication of the Virus Genome

Included in Steps III and IV are the specific components of the viral biosynthetic machinery, such as specific enzymes. There are many examples of novel antivirals in this area. For example, new approaches to the inhibition of the HIV protease are likely to reach clinical trials shortly. In this book, Merluzzi et al., report an exciting novel small organic compound characterized by a cyclic propyl group. This agent, called BI-RG-587, has potent activity against HIV-1 reverse transcriptase, but little against the HIV-2 enzyme. To our mind, this is the first example of a non-nucleoside inhibitor of reverse transcriptase and is unusual in its profound selectivity. This compound was the focus of a great deal of press attention in December, 1990, shortly after our meeting convened, and we are pleased to include a detailed description of that work here.

Rossi and Sarver describe the construction of ribozymes that cleave HIV-specific RNA. In another contribution, Cantin et al. show that antisense oligonucleotides can be used to inhibit herpes simplex virus (HSV) growth in tissue culture. They even offer speculation as to how antisense oligonucleotides might be delivered by retrograde transport to ganglion that latently harbors HSV. Although the examples used are for HIV and HSV, respectively, it is clear that other viruses can be targeted with antisense and catalytic RNA designs. Both systems exploit the specificity conferred by genomic sequences and avoid the necessity of understanding specific gene product function.

Specific delivery to liver tissue was considered by Hsieh and Taylor, who propose designs for a disarmed hepatitis delta virus that will specify a ribozyme with activity against hepatitis B virus (HBV). Ideally, this hypothetical "attack" virus would land harmlessly in uninfected liver cells. However, since HBV and Delta virus infect the same cells, HBV in cells receiving the attack delta virus, would be interfered with. Although the notion of deliberate release of a delta virus into patients already chronically infected with HBV in unsettling, the idea illustrates how a virus with a particular narrow tissue tropism can be used as a specific delivery mechanism. Other full-length contributions that discuss targeting specific virus enzymes are by Ho et al., and Cheng et al.

Step VII: Immune Modulations

Martin et al. discuss the use of alpha interferon in the treatment of viral hepatitis. Gerin et al. take this to the next step and discuss the use of thymosin alpha peptide, a completely new development in the immune modulator therapy of viral hepatitis.

Detection of Virus Infections

This book also reviews the major advances and frustrations associated with virus detection systems. Dr. Thomas Smith discussed the rapid detection of viruses using improved methods for growth (dram vials) and improved immunological methods for detection of viruses. ELISA detection methodologies haved lived up to their promise when the number of viral serotypes is limited and the number of infecting particles is high. In earlier years, the use of DNA probes held much promise and were touted as the successor to culture for the rapid, sensitive, and specific detection of viral infections. Dr. Harley Rotbart discussed DNA probes and emphasized their somewhat disappointing results in meeting the sensitivity requirements for laboratory diagnosis. The overall summary of the series of papers on diagnostic techniques was that as more antigens are purified and as the good nucleic acid PCR primers are selected, the goal of rapid and sensitive detection of viral infection will be realized.

Antivirals from the Sea and Isolation of Antivirals from Rationally Designed Mutant Bacteria

Finally, two completely different ways to search and screen for antiviral agents are presented. Kenneth Rinehart describes trips into the ocean to collect samples that are tested for antiviral activity on board. These adventures have resulted in the discovery of several exciting natural products. At least one class, the didemnins, isolated from a sea tunicate, has significant activity against the herpesviruses and is currently undergoing clinical evaluation in phase II trials (for possible cancer chemotherapy).

The idea of looking to natural products for chemotherapeutics is brought back into the laboratory in the chapter by Grafstrom et al. This section discusses the possibility of genetically engineering bacteria (or other microorganisms) so that mutants can be selected that may themselves be capable of producing antiviral agents. That is, bacteria have been made that express an important viral function, such as the HIV protease. Bacteria have been constructed such that they can only grow in a selective media if the important viral function (protease) is inhibited. Mutant bacteria that have acquired the ability to grow are isolated and tested for the possibility that they produce an inhibitor of the protease. Such a system could have profound implications, since it is relatively easy to generate bacterial mutants. To our mind, this is the first example where genetic selection has been applied to bacteria to search for antiviral agents.

In addition, the book contains other highly readable and important contributions of a more specialized nature. For example, Dr. Cindy Wordell, Dr. Donald Jungkind et al., presented a paper on how to assay for anti-CMV activity in intravenous gamma globulin preparations. This *in-vitro* method will be useful to gauge potential potency differences between lots and brands of intravenous gamma globulin products. We are confident that anyone interested in learning about new approaches to antiviral development and the detection of viruses will find a great deal of useful information within these bindings.

Timothy M. Block

Donald L. Jungkind

Richard Crowell

Mark Denison

REFERENCES

1. Enders, J.F., Weller, T.H., and Robbins, F.C., Cultivation of the Lansing strain of poliomyelitis virus in cultures of various human embryonic tissues, Science 109:83087 (1949).
2. Pasteur, L., Chamberland, C., and Roux, E., Physiologic experimental nouvelle communication sor le range, C.R. Acad. Sci. 98:457-463 (1884).
3. Reed, W., Recent researches concerning the etiology, propogation and prevention of yellow fever by the United States Army Commission, J. Hyg. 2:101-119 (1902).
4. Rous, R., Transmission of a malignant new growth by means of a cell free filtrate, JAMA 56:198 (1911).

CONTENTS

STUDIES OF REOVIRUS PATHOGENESIS REVEAL POTENTIAL SITES FOR ANTIVIRAL INTERVENTION

Bernard N. Fields
Harvard Medical School

SUMMARY

Pathogenesis studies in animals can uncover details concerning viral replication, growth, and access to target organs, *in vivo*. This, in turn, reveals opportunities for antiviral intervention that may be otherwise missed by limiting analysis to growth of virus in tissue culture. In this report, reovirus infection of mice is used as a model. Three general aspects of reovirus behavior in mice are presented and each demonstrates a property of the virus that could easily have been missed by studies in tissue culture.

INTRODUCTION

This report will emphasize the importance of studying the behavior of viruses within their natural hosts. That is, although most recent experiments have been done using cell culture systems in which numerous important and remarkable discoveries have been made, there are properties and viral functions that may be overlooked if analysis is limited to tissue culture systems. With that in mind, it will be easy to see how pathogenesis studies can be used in the pursuit of antiviral agents. Moreover, this report will show several examples of steps in the reovirus life cycle within its host that may be useful in fashioning antiviral intervention. These properties fall into two general categories; host modification of whole virus during passage through the gut and post attachment movement of the virus through the Central Nervous System (CNS). Briefly, the observation that reovirus is proteolytically modified or promoted to infectivity in the mouse gut will be reviewed. In addition, the nature of how reovirus spreads throughout its infected host and can have its access to nervous tissue prevented by post inoculation treatment with antibody, will also be considered. Both properties would have been missed had analysis been limited to tissue culture.

BACKGROUND

The focus of this presentation is on reovirus. However, the broader purpose is to discuss pathogenesis. By that I mean the relationship between a virus and its host. In that sense, reovirus serves as a general model for many other viruses. This is legitimate because thousands of viruses use common pathways and many of the principles invoked to explain and dissect out the reovirus pathogenesis are common to many viruses. At each point I will try to generalize from the specific example provided by reovirus.

For example, taking the virus-host interaction step by step, consider the following. First, the virus must enter its host. There are not many places that a virus can enter its host. The mouth, respiratory, and genital tracts are common natural routes. Systemic entry can occur from blood transfusion or tissue transplantation. Alternatively, viruses can be introduced experimentally by injection into muscle and blood vessels. For reovirus, the virus is swallowed. Following this, the virus must find its way into the site of its primary replication. After swallowing, reovirus enters the small intestine. Therefore, not that prior to entering a cell, reovirus has travelled a remarkable distance in its host. Similar to other enteric viruses, it must survive an extraordinarily harsh environment, including bile salts and proteases. Since this hostile environment is encountered by all enteric viruses, it is not surprising that they share many common structural features and use many common survival strategies. This, of course, reinforces the point that by studying one virus (such as reovirus), we will learn a great deal about the others.

Following passage into the gastrointestinal tract, reovirus arrives at barrier cells that number only a few hundred per mouse and cannot be cultured *in vitro*. These are specialized intestinal epithelial cells that overlay the dome of Peyer's patch. They are membranous and full of microfolds, hence are called "M" cells. They transport reovirus (as well as poliovirus and many bacteria) across the lumen and into macrophages of the Peyer's patch. The attachment is to luminal (apical) M cell surfaces. Movement into intracellular vesicles and the intercellular space between M and mononuclear cell surfaces has been observed using electronmicroscopy. M cells, therefore, act to transport virus from one surface to another, without grossly modifying the virus. This has also been confirmed by analysis of radiolabeled virus, recovered following passage through the stomach and into the small intestine.

In macrophages in the Peyer's patch, the virus now undergoes primary replication and spreads to many of the secondary organs shown in Figure 1. Reovirus access to the brain provides an excellent model of the pathogenesis of a neurotropic virus. The numerous insights already provided by studying the reovirus model have been amply reviewed elsewhere and will not be repeated here. Reviews of this can be found in references 4, 5, and 6. Rather, this report will concentrate on the three aspects referred to in the introductory paragraph. Parenthetically, the immune system is responding to the infection with its complex machinery. Unfortunately, space does not permit consideration of the immune response to virus infection, and for that the reader is encouraged to look elsewhere.

Therefore, this introduction presents three aspects of virus behavior in the host that cannot be completely studied or reproduced in tissue culture. First, the long trip the virus makes following entry and access to the lumen; second, events that occur to help spread virus to secondary sites of the nervous system; and third, the growth and movement of virus through the nervous system in the face of (fourth) a mounting immunological response.

HOST STRUCTURAL MODIFICATION OF REOVIRUS

The mammalian reoviruses are 70 nM in diameter, non-enveloped and contain 10 segments of ds RNA as their genome. There are three reovirus gene products that most directly influence its pathogenesis: μ1C, σ1 and σ3. These are the outer capsid proteins. The reovirus σ3 protein is the 41,000 Dalton polypeptide product of the S4 gene. There are approximately 720 σ3 and μ1C polypeptides located on the viruses outer capsid (Figure 2). σ3 is considered to be "functionally" equivalent to a "spore" protein, since it confers stability. It has a zinc binding domain analogous to cellular

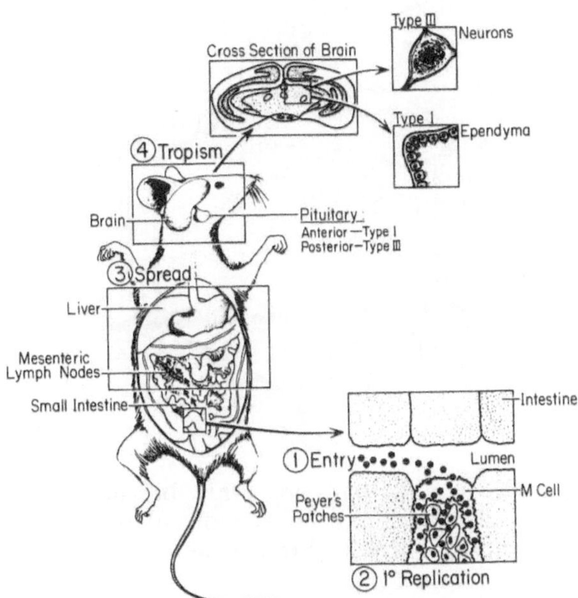

Figure 1. Steps in the entry and spread of reovirus in the mouse. Each number represents a discreet "step" in the life cycle of the virus, starting with 1, entry (the virus arrives at the intestine following oral entry) and ending with 4, tropism (which shows the virus arriving in the CNS) (From: Reference 6)

transcription factors, such as TFIIIA, and presumably serves to bind zinc to the virus outer surface. In addition to structural roles, genetic studies suggest that σ3 plays an important role in regulating host macromolecular synthesis. Strains of reovirus that establish persistent infections have mutations within this gene. In addition, transcription and translation of virus genetic information is influenced by σ3.

μ1C is the 72,000/80,000 (proteolytically processed) Dalton polypeptide product of the M2 gene. Its proteolytic processing results in transcriptase activation. Moreover, it has been associated with influencing the neurovirulence of the virus.

σ1 is the 38,000 Dalton product of the S1 gene. It is also an outer capsid protein and is responsible for virus attachment to the cell as well as the characteristic hemagglutination properties of the virus. Since the pathogenicity of reoviruses depend, to some extent, upon their ability to survive the passage through the gut, it has been proposed that σ1 (and other outer capsid proteins) play an important role in determining this stage of pathogenesis. Indeed, recent studies have supported this hypothesis. Genetic analysis of the differences between reovirus type 1 (Lang), which grows well in intestinal tissue and reovirus type 3 (Dearing), which does not, have revealed that most of the ability of the virus to survive the gut can be attributed to σ1 (and the L2 gene product, λ2, a virus capsid protein) (2).

When reovirus encounters host proteases, σ3 is processed (proteolytically cleaved), causing it to unfold and possibly activate an intrinsic protease viral property. Activated σ3 may cleave μ1c, the other abundant outer capsid protein, into a lower molecular form, called delta(δ). I will provide evidence that these effects are physiologically important in enhancing the ability of the virus, in the processed form, to attach to its target cells.

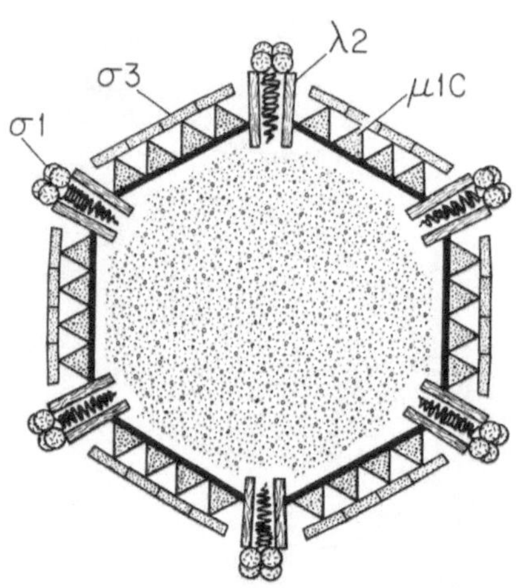

Figure 2. Schematic of a reovirus particle. Only the capsid proteins are shown. Note: μ1c is cleaved into a lower molecular weight form called delta (δ). See test and Figure 3. (From Reference 4)

In one set of experiments, reovirus was purified from tissue culture, following labeling with ^{35}-S-methionine. Radiolabeled virus was introduced directly into the mouse stomach. As shown in Figure 3A, following introduction into the stomach, σ3 is rapidly cleaved. This is followed by the processing of μ1c into its δ form.

The effect of the proteolytic processing upon the virion particle has been followed by electron microscopy. Samples analogous to those used in the SDS-polyacrylamide gel electrophoresis experiments were imaged with the electron microscope (Figure 3B). It can be seen that the proteolytic removal of sigma 3 results in a major conformational change in the virion. The virus appears to assume a more icosahedral shape, and has extended the attachment protein σ1, allowing for improved access to target sites on the host cell. Note that all of this takes place in the absence of a cellular environment.

The proteolytic modification of the virion, as shown in Figure 3B. results in a virus particle called the Intermediate Subviral Particle (ISVP). It has a high affinity for target cells, but is less stable than the native, unprocessed virion. We believe that it is the virus primarily in the form of an ISVP that attaches to the M cells and is transferred through the Peyers patch. Therefore, it is the ISVP that is delivered to the macrophages, where primary replication ultimately occurs.

We next asked if generation of an ISVP by proteolysis is essential for efficient viral access to target organs and subsequent replication. That is, if interruption of the processing of reovirus (or other enteric viruses) interferes with the virus life cycle in its host, a vulnerable part of its life cycle will be recognized and made available for antiviral intervention. To test the question of the role of ISVP formation in the virus life cycle in the mouse, my colleagues and I conducted experiments using protease inhibitors (1). In control experiments, radiolabeled intact virus (Figure 4, lane 1) was treated with chymostatin and converted, *in vitro*, into the ISVP form, as evidenced by the diminution of μ1c and the appearance of δ (Figure 4, lane 2). Alternatively, intact radiolabeled virus was introduced into the stomach of a mouse that had been given

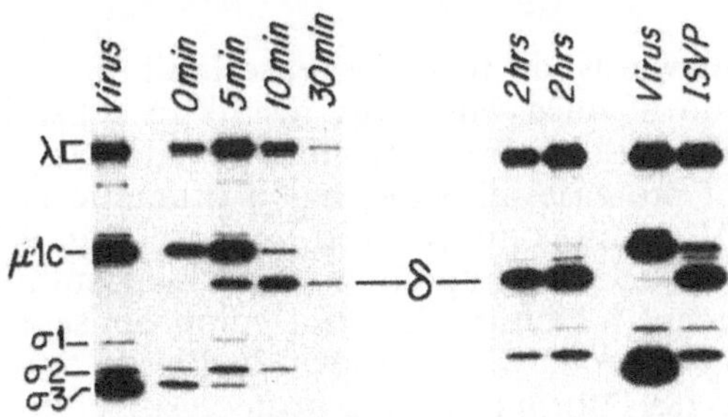

Figure 3A. Proteolytic processing of the reovirus virion A) Fluorogram showing the effect of *in vitro* digestion of virus by various gut proteases. ^{35}S-labeled virus was digested with 1 mg of protease per ml for 1 hour at 37°C and then diluted in Laemmli sample buffer and subjected to SDS-PAGE. Lane 1: pepsin digest, 2: mock digest, 3: chymotrypsin digest, 4: trypsin digest, 5: virus recovered from the small intestine of a suckling mouse 2 hours after peroral inoculation. Position of the proteins δ, μ1c, σ and σ3. 3A. From: Reference 3.

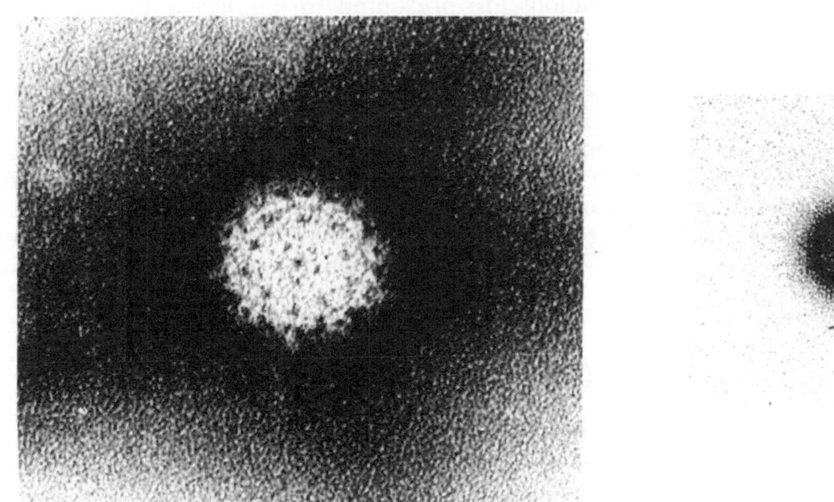

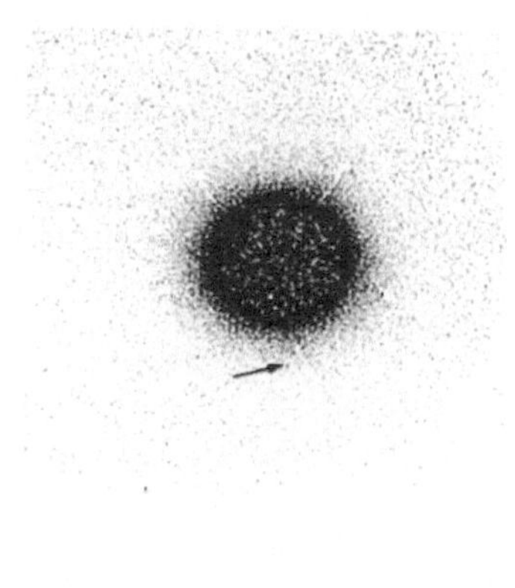

Figure 3B. Electron micrograph of a reovirus particle before (left panel) and after (right panel) proteolysis. Note that σ1 protein appears "extended" in virus that has been digested with protease (arrows). 3B. courtesy of D. Furlong.

neither aprotinin or chymostatin alone, one hour prior to infection. Subsequent recovery of virus shows little inhibition of the ability of mouse proteases to convert the virion into an ISVP (Figure 4, lanes 4, 7, and 8). However, lanes 5 and 6 (Figure 4) show that pretreatment of mice with both chymostatin or aprotinin, in combination, prior to challenge with labeled virus results in a substantial degree of interference of the conversion of virus into the ISVP form.

Having shown that the conversion of virus to the ISVP can be prevented by combination treatment with protease inhibitors, we now asked if virus, frozen in the pre-ISVP (native) form had an altered ability to infect and replicate in target cells in the mouse. In other words, do these protease inhibitors have an antiviral effect *in vivo*? First, infection of mice with virus results in substantial (wildtype) levels of replication in intestinal tissue within 24 hours (Figure 5). The presence of a single combination dose of protease inhibitors has a dramatic effect on the ability of the virus to replicate. Virus titers are reduced by more than 90% compared to controls when mice are pretreated with aprotinin and chymostatin (Figure 5). Note that treatment with protease inhibitors does not compromise the ability of the mice to replicate reovirus, in general. This is shown in Figure 5, where virus that has been pre-converted to the ISVP form by *in vitro* proteolysis replicates as well as in protease inhibitor treated mice as in untreated mice. Therefore, protease inhibitors do not, per se, have an impact upon primary viral replication.

In other experiments, it has been shown that continuous infusion of protease inhibitors resulted in an even more dramatic reduction in the primary viral replication in the mouse. There is at least a 100 fold reduction in virus replication. Moreover, the ability of precleaved virus (ISVPs) to replicate is also affected. That is, mice that are continuously provided with protease inhibitors and infected with ISVPs initially replicate virus normally. However, subsequent rounds of replication (secondary rounds) are significantly inhibited. This suggests that reinfection beyond the Peyer's patch step also requires conversion of progeny virus into an ISVP.

These experiments and observations are presented to emphasize the point that important events in the life cycle of a virus occur prior to its entry into a cell. The proteolytic processing described here occurs in the intestinal lumen, well after its entry into the host and well before entry into the hospitable environment of a cell. Similar results concerning proteolytic processing/maturation have been observed for another member of the reovirus family; notably rotavirus, by Robert Yolken and colleagues (10). Moreover, the phenomenon of proteolytic processing of viruses also occurs for many other virus families, such as influenza, sendai and other paramyxoviruses. These steps may therefore provide valuable opportunities for intervention and may not have been fully appreciated in the case of reovirus without pathogenesis studies in the live animal.

VIRUS SPREAD WITHIN ITS HOST

The next topic that will be considered in this presentation is the movement of reovirus, in particular, and neurotropic viruses, in general, throughout the nervous system. For example, if virus is introduced into the footpad of a mouse, how does it move from the site of inoculation to the site of primary replication and ultimately throughout the host. This is an important question since many viruses do not cause disease at their point of entry or even at the point of primary replication. Preventing the spread or dissemination of virus throughout is host may therefore present another opportunity to prevent its ability to cause overt disease. Therefore, virus spread throughout its host represents another important aspect of pathogenesis which cannot be studied well in tissue culture.

Consider two possible means of virus access to the nervous system following footpad inoculation. In the first model, following entry into the footpad, virus directly attaches to and moves through the sciatic nerve, rising to the inferior and then superior spinal cord. In the second model, viremia is necessary prior to nervous access and, following footpad injection, virus replicates in muscle, disseminates throughout the blood and arrives in many parts of the spinal cord at the same time. These two strategies can be distinguished experimentally and the strategy reovirus uses, for example, can be determined. Briefly, the footpads of mice can be inoculated with

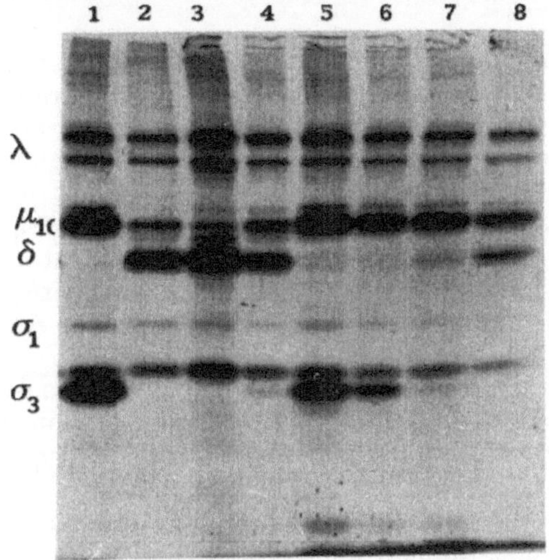

Figure 4. Fluorogram showing the effect of protease inhibitors on *in vivo* virus digestion in the intestinal lumen. Radiolabeled virus was administered to 10-day-old suckling mice which had been pretreated with phosphate-buffered saline or protease inhibitors. Mice were killed 4 hours post-inoculation, and virus was recovered from the intestinal lumen by rinsing and was then subjected to SDS-PAGE and fluorography. Lane 1: ^{35}S-labeled purified type 1 reovirus, 2: virus digested with 0.1 mg of chymotrypsin per ml *in vitro*, 3: virus recovered from a mouse pretreated with phosphate-buffered saline, 4: virus recovered from a mouse pretreated with aprotinin alone, 5 and 6: virus recovered from two mice pretreated with aprotinin and chymostatin, 7 and 8: virus recovered from two mice pretreated with chymostatin alone. Position of the proteins A) μ1c, B) σ1 and σ3 are indicated. From Figure 2, Journal of Virology, 1990, 64:1830 (From Reference 1).

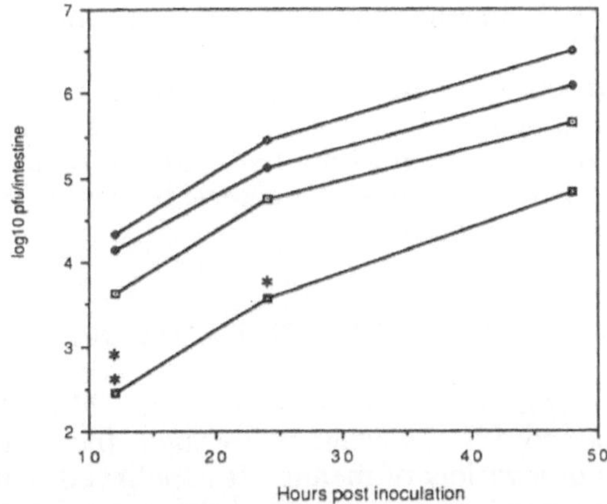

Figure 5. Viral titers in intestinal tissue after a single dose of protease inhibitors. Mice were inoculated with 10 PFU of type 1 reovirus (square and square) or type 1 reovirus ISVPs 1 hour after pretreatment with either phosphate-buffered saline or protease inhibitors. Each data point represents the mean of three to five animals. Symbols: < and # indicate P <0.05 and P<0.02 compared with phosphate-buffered saline-pretreated type 1 reovirus respectively (From Reference 1).

reovirus and virus can be recovered from different regions of the spinal cord at a given time, post infection. If the second strategy is used (and viremia is necessary for access to the spinal cord), virus should simultaneously appear at different regions of the spinal cord. On the other hand, if the first strategy is used and there is a sequential, heirarchical, pathway of spread of virus from footpad, to sciatic nerve, etc., the appearance of virus at each point in the pathway will be temporally distinct and occur in a predictable order. It is believed, parenthetically, that herpes simplex and rabies viruses use the first strategy and most enteroviruses, arboviruses, measles virus, mumps virus, and lymphocytic chorio-meningitis virus requires a viremic phase.

Reovirus provides valuable genetic tools that can demonstrate the effectiveness of the mouse model in answering these pathogenesis questions (7). Different strains of reovirus appear to use different pathogenesis strategies. Type III Dearing reovirus has been shown to follow the strategy outlined in model one, gaining direct access to nervous tissue without viremia, following footpad injection. However, the Lang strain (type 1) of reovirus does appear to use a blood phase for nervous access. This hypothesis was tested in the mouse model and Figure 6 shows the kinetics of virus appearance in the inferior and superior spinal cord as a function of time following hind limb inoculation. Notice that virus appears in the inferior and superior spinal cord simultaneously, following hind limb injection of type 1 reovirus (Lang strain). Similar results are observed if this strain is injected into the fore limb. Moreover, a similar pattern of simultaneous appearance of virus in inferior and superior spinal cord regions is seen even if the sciatic nerve is severed prior to inoculation of hind limb. This suggests that access from the hind limb to deeper nervous areas does not necessarily use the sciatic nerve pathway but is spread in the blood.

A different pattern of kinetics is seen with Reovirus type III (Dearing). Following injection into the fore limb, viruses appear first in the superior spinal cord and only later in the inferior cord. Following injection into the hind limb virus appears first in the inferior cord and not until three days later in the superior spinal cord. This shows a sequenced pattern of movement, dictated by the relationship of the nervous pathways relative to the site of injection.

The effect of surgical interruption is a critical experiment and is shown in panels "C" of Figure 6 and 7. In this experiment, severing the sciatic nerve prior to injection into the hind limb prevents type III Dearing but not type I Lang virus from appearing in the spinal cord. This shows that an ordered continuum between injection site and spinal nerves cannot be interrupted without affecting reovirus type III access to distant organs, but reovirus type I Lang virus access is sciatic nerve independent (and presumably is dependent on viremia).

MECHANISM OF THE SPREAD OF A NEUROTROPIC VIRUS

As shown, reovirus type III Dearing is neurotropic. How does it move through the nerves of the spinal cord? Movement through the nervous system could be accomplished by any of a variety of means. It is believed that reovirus travels within an axon using fast transport. Early support of this came from an observation by Sam Dales in 1965 (reviewed in 4). He showed that in infected cultured mouse L cells, reovirus is associated with microtubules. Consistent with this, we have shown that reovirus actually binds to purified microtubules, *in vitro*, with high affinity. Back in 1965, the significance of microtubule association was not clear. Moreover, since the addition of colchicine to reovirus infected cells, in tissue culture, did not influence viral titers, the relevance of observing the virus' association with microtubule structures was

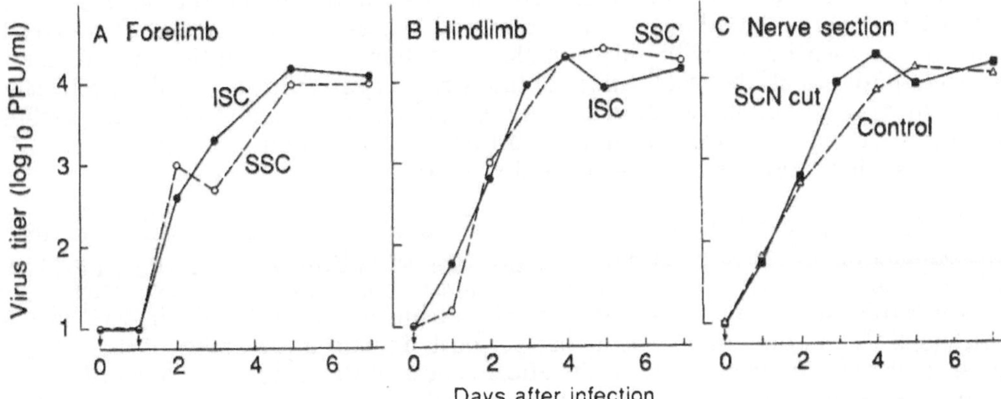

Figure 6. Pattern of spread of reovirus type 1 to the spinal cord of neonatal mice after inoculation of virus into the fore limb (A) or hind limb (B) footpad. (O) Superior spinal cord (SSC); (O) inferior spinal cord (ISC). Each plotted point represents the average log 10 titer of five to seven specimens. The SEM is less than 0.6 for SSC and ISC at all time points. (C) Spread of type 1 to the ISC after sectioning of sciatic nerve (square) is compared with spread in control animals with intact nerves (triangle). A downward arrow indicated no virus detected at lowest dilution shown (From Reference 1).

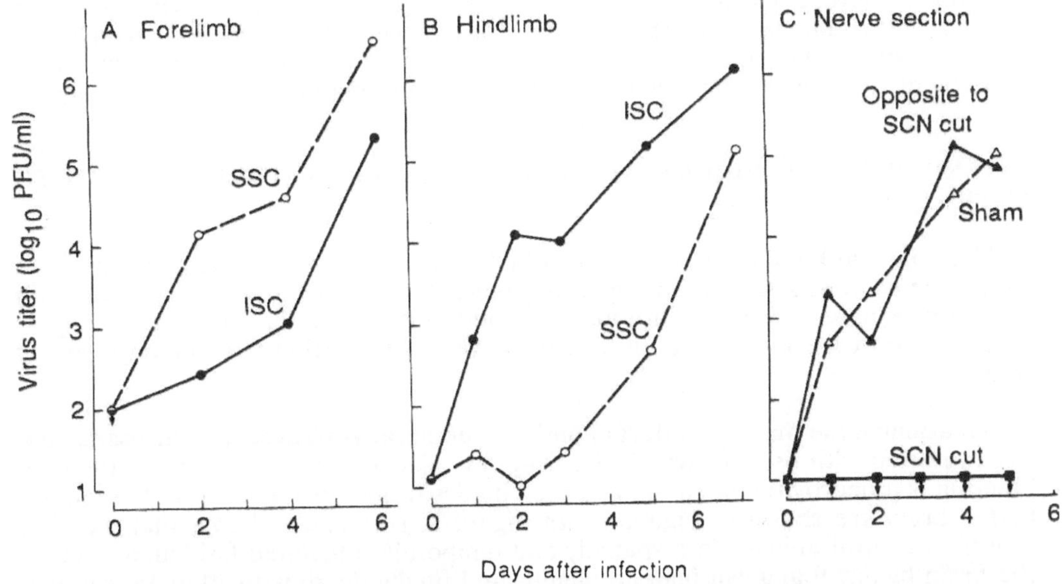

Figure 7. Pattern of type 3 reovirus spread into the spinal cord after inoculation into either fore limb (A) or hind limb (B) footpad. Symbols (O), superior spinal cord (SSC); (O), inferior spinal cord (ISC); Panel C: Sciatic nerve sectioned in limb ipsilateral (square) or contralateral (triangle), to virus inoculation. (▲) sham operation. No virus detected at the lowest dilution shown (From Reference 7).

uncertain. However, microtubules form the essential components of fast axonal transport in nerves. Perhaps reovirus is moved via fast axonal transport in nerves by way of its association with microtubules. In this case, colchicine would not be expected to affect the replication of the virus in cell culture. Rather, colchicine, through its affect on microtubule formation, would be expected to inhibit the mobility or transport of the virus from two distinct points within nervous tissue. This hypothesis has been tested experimentally, as follows (7).

Following colchicine inoculation into a mouse hind limb, reovirus Dearing is introduced at the same location. A single dose of colchicine delays the appearance of virus in the inferior spinal cord by at least three days, suggesting microtubule assisted transport of the virus from the site of inoculation was inhibited (Figure 8). Although this amount of colchicine had seriously impaired the ability of the mouse to walk, giving dramatic evidence of the effectiveness of the drug, there was little impact upon the replication of the virus at the inoculation site, compared with untreated control mice.

The experiments described in the preceding paragraph show that colchicine prevents access of the virus to various regions of the nervous system. Perhaps even more compelling is the evidence that colchicine prevents spread of the virus throughout the nervous system, even after initial entry. Mice were inoculated with reovirus, as before, and "post treated" with colchicine at 24 hours afterwards. Since reovirus lethality is associated with infection of higher centers (and not the spinal cord) this is evidence that virus spread to these centers has been prevented, even after initial replication of the virus and entry into the nervous system has begun. Therefore, colchicine is not blocking multiplication at the primary site of entry. Rather, it is blocking its spread through nervous transport systems, and in this sense provides another vulnerable point in the life cycle of the virus that may be amenable to antiviral agents. This is, incidentally, another example of virus behavior that would be difficult to recognize by strictly *in vitro* (tissue culture) analysis.

INTERRUPTION OF VIRUS SPREAD BY POST INFECTION TREATMENT WITH ANTIBODIES

The role that humoral immunity plays in limiting the spread of a neurotropic virus within the nervous system is an important topic. It is particularly important that this subject be examined *in vivo*, since many antibodies that score as "non-neutralizing" in tissue culture experiments are, in fact, effective in influencing virus pathogenesis *in vivo*.

To examine the matter of effect of antibody upon reovirus access to the brain, mice were infected with reovirus type 3 (Dearing) by injection into the footpad. Reovirus Dearing is a neurovirulent strain, and the kinetics of its movement from the footpad to the brain are shown in Figure 7. In Figure 9, it is again shown, that the virus moves in control animals in a spatially and temporally regulated fashion, arriving in the brain by day 9 and causing encephalitis and finally death with 10 to 15 days post inoculation.

However, the story with mice inoculated in the footpad in the presence of antibody is quite different. Figure 9 shows that virus titers in the ISC, SSC and brain are greatly suppressed and mice are completely protected from the lethal effects of reovirus. In a sense this is not too surprising since the monoclonal antibody (G5) is neutralizing in nature.

More surprising is the impact that antibody administered following infection has upon virus spread. Following the fate of virus injected into the muscle of the footpad (predominant consequence for virus following footpad inoculation) shows most of the virus present within muscle tissue, within the first 24 hours (not shown). There is little, if any, effect when this infection is followed by either control (non-neutralizing) or neutralizing antibody (Figure 10). However, in animals post treated (24 hours after infection) with the neutralizing antibody, very little virus reaches the superior spinal cord, compared to those treated with control antibody. There has been little effect upon primary replication in the muscle of the footpad (primary site) and virus reaches the inferior cord, but transport to the superior cord is completely blocked. Thus when antibody and virus are added via the peripheral route, access of the virus to the CNS is prevented, even when these two contributions are temporally separated by 24 hours.

To what extent does post infection administration of antibody paralyze the virus and prevent its spread through neural tracts? This was dramatically answered by providing antibody intraperitoneally to mice that have been injected intracerebrally with at least 100 lethal dose 50s (LD_{50}) and isolating virus from various tissues. Although initially, virus titers in the brain are high, virus spread throughout the CNS as monitored by recovery of the virus in the eye, and other connected tissue, is completely inhibited. Figure 11 also shows these results. Moreover peripherally administered antibody completely protects the mice from the lethality of reovirus encephalitis, even though the virus was introduced directly into the CNS (not shown). This is very impressive protection. This shows that antibody protects mice from reovirus replication in the CNS and lethality, even after substantial access to the CNS was achieved.

The impact of peripherally administered antibody (intraperitoneal) upon intracerebrally inoculated virus is complex. In the absence of antibody, virus moves through neurons innervating the optic nerve and orbit. This is completely blocked by the antibody. In addition, there are tracts of nerves within the brain, and antibody clearly prevents movement of virus within certain tracts. This is the critical point, because the brain is where the most serious pathology is caused by reovirus and it is the interruption of the virus access to these specialized regions within the brain that must be protective. That is, survival of the mouse host permitted by antibody, must be due to the sparing of these higher order centers. Parenthetically, virus can provide a novel way to study the communication of these tracts with each other.

The above results were obtained with the virus introduced via the footpad. Similar information can be obtained using virus introduced into the gut parenterally. Virus grows in the Peyer's patch equally well with or without post administration of antibody. Post administration of antibody does prevent, however, access of the virus to the inferior and superior spinal cord. It is now known that reovirus travels from the Peyer's patch through autonomic nerves, retrograde, through parasympathetic fibers to the vagal nucleus.

These models show that for viruses in which there is a period between entry into the host and primary replication and minor illness and major illness, it may be possible to purposefully administer antiviral agents that inhibit movement of the virus. This model is valid for all of the enteroviruses. The antiviral might be a specific antibody or other pharmacologically active compound. Conventional concerns about crossing the blood brain barrier was not a limitation for the antibody mediated effect on the CNS in the experiments described here. This alone raises some interesting possibilities concerning the kinetics and location of antiviral administration.

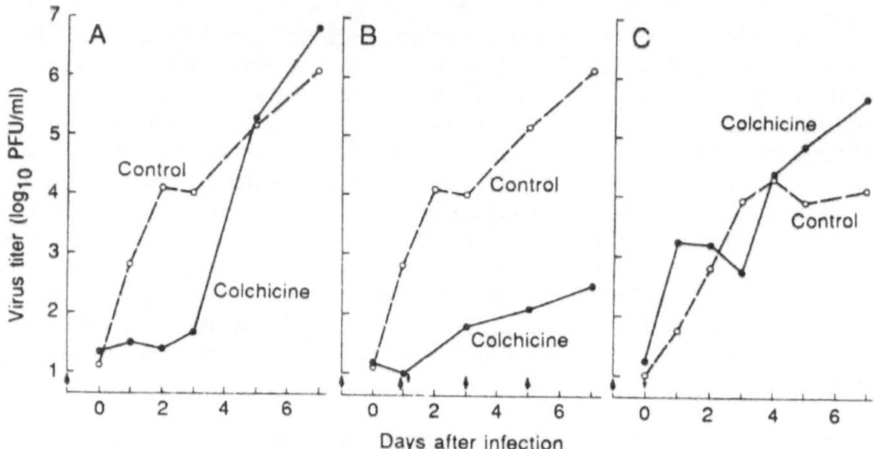

Figure 8. Pattern of spread of type 3(A and B) or Type 1(C) to the inferior spinal cord in colchicine-treated (O) and untreated (O) mice. Upward-pointing arrows above the time scale indicate time of colchicine administration. Colchicine (J.T. Baker)-treated mice received 2 x 10^{-7} g of colchicine per gram of body weight. For studies of the effect of a single dose of colchicine (A), 1-day-old mice were given 0.010 ml of 5 x 10^{-5} M (20 x 10^{-6} g/ml) solution of colchicine into a hind limb muscle by means of a 30-gauge needle and a Hamilton microsyringe. In (B) animals received 2x 10^{-7} g of colchincine per gram of body weight; either a 5 x 10^{-5} M colchicine solution (ages 1 and 3 days) or a 1 x 10^{-4} M solution (ages 5 and 7 days) was used. Only mice showing evidence of hindlimb analgesia and paresis were used for the colchicine multidose experiments. Virus was inoculasted in a 0.010- to 0.015-ml volume into the hindlimb footpad 24 hours after the first colchicine dose or at age 2 days (controls). The dose of virus was 3.8 x 10^5 to 4.5 x 10^5 PFU. For colchicine-treated animals, each plotted point represents the average of the log_{10} titer (plaque-forming units per milliliter) of ISC specimens from two to four animals (A and C) or five to seven animals (B). (From Reference 7).

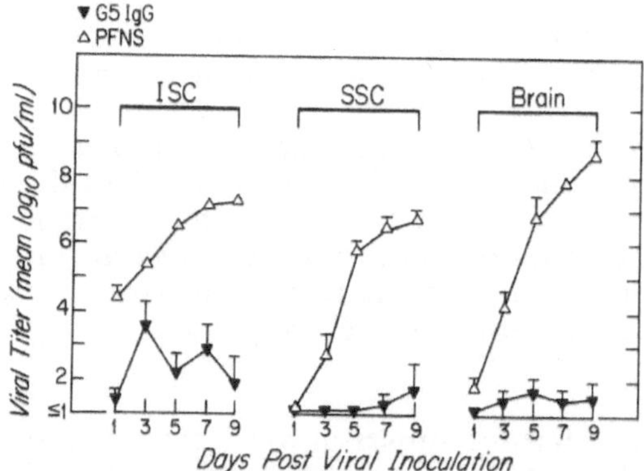

Figure 9. Effect of antibody administered at the time of viral inoculation. Mice were given either sigma 1mAbG5 (▽) or saline (△) intraperitoneally (IP). At the same time, mice were inoculated with T3D in the hind limb. At the indicated times postinfection samples of ISC, SSC, and brain were collected and assayed for virus (five to seven mice per data point). Only upward error bars are shown for clarity (From Reference 8).

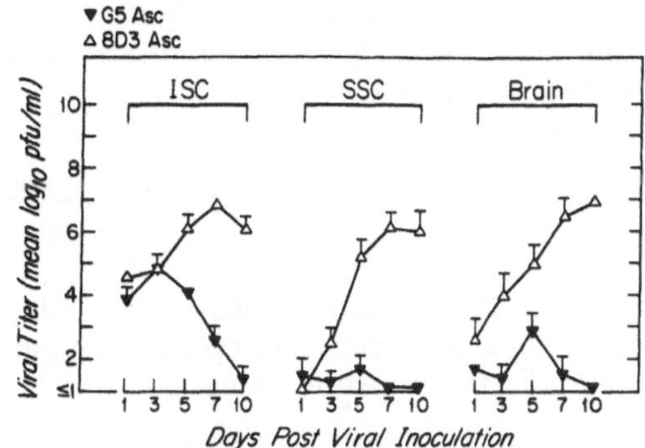

Figure 10. Effect of antibody administered after viral inoculation. Mice were inoculated in the hind limb footpad with T3D. 24 hours later mice were give s1mAbG5-containing(▽) or control (8D3)(△) ascites. At the indicated times, samples of ISC, SSC, and brain were collected and assayed for virus (two to four mice per data point). Only upward error bars are shown for clarity (From Reference 8).

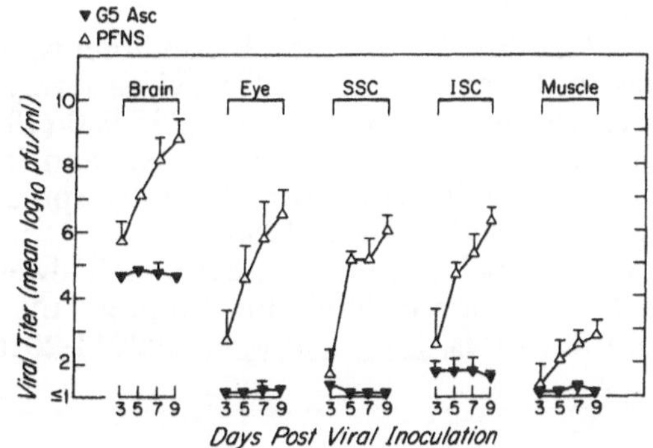

Figure 11. Effect of antibody administered after intracerebral (IC) inoculation. Mice were inoculated with 3 log_{10} PFU of T3D IC. 24 hours later mice received either σ 1mAbG5-containing ascites(▽) or PFNS IP(△). At the indicated times, samples of brain, eye, ISC, SSC, and muscle were collected and assayed for virus (controls: two to three mice per data point; antibody-treated: five to seven mice per data point). Only upward error bars are shown for clarity (From Reference 8).

CONCLUSION

Three general aspects of virus behavior in its natural host have been presented that provide opportunities for intervention. These aspects each represent a viral or host property that could not be easily studied in tissue culture systems. This presentation therefore underscores the importance of balancing *in vitro* models of virus molecular biology with *in vivo* analysis of pathogenesis.

REFERENCES

1. Bass, D.M., Bodkin, D., Dambrauskas, R., Trier, J.S., Fields,B., and Wolf, J.L.,Intraluminal proteolytic activation plays an important role in replication of type 1 reovirus in the intestine of neonatal mice, J. Virol. 64:1830-1833 (1990).
2. Bodkin, D.K., and Fields, B.N., Growth and survival of reovirus in intestinal tissue. Role of the L2 and S1 genes, J. Virol. 63:1188-1193 (1989).
3. Bodkin, D.K., Nibert, M.L., and Fields, B.N., Proteolytic digestion of reovirus in the intestinal lumens of neonatal mice, J. Virol. 63:4676-4681 (1989).
4. Schiff, L.A., and Fields, B.N., "Reoviruses and their replication," Chapter 47, pp. 1275-1305, in: "Virology, Volume 2", B.N. Fields, et al., eds., Raven Press, New York, 1990.
5. Tyler, K.L., and Fields, B.N., Reoviridae: A brief introduction, Chapter 46 in: "Virology, Volume 2," B.N. Fields, et al., eds., Raven Press, New York (1990).
6. Tyler, K.L., and Fields, B.N., et. al., Reoviruses, Chapter 48, in: "Virology, Volume 2," B.N. Fields, et. al., eds., Raven Press, New York (1990).
7. Tyler, K.L., McPhee, D.A., and Fields, B.N., Distinct pathways of viral spread in the host determined by reovirus S1 gene segment, Science 233:7709-7774 (1986).
8. Tyler, K.L., Virgin, H.W., IVth, Bassel-Duby, R., and Fields, B.N., Antibody inhibits defined stages in the pathogenesis of reovirus serotype 3 infection of the central nervous system, J. Exp. Med. 170:887-900 (1989).
9. Virgin, H.W., IVth, Bassel-Duby, R., Fields, B.N., and Tyler, K.L., Antibody protects against lethal infection with neurally spreading reovirus type 3 (Dearing), J. Virol. 62:4594-4604 (1988).
10. Vonderfecht, S.L., Miskuff, R.L., Wee, S.B., Sato, S., Tidwell, R.R., Geratz, J.D., and Yolken, R.H., Protease inhibitors suppress the in vitro and in vivo replication of rotaviruses, J. Clin. Invest. 85:2011-2016 (1988).

DRUGS AS MOLECULAR TOOLS

Frank J. Dutko[#], Donald E. Baright[#], Guy D. Diana*, M. Pat Fox[#], Daniel C. Pevear[#], and Mark A. McKinlay[#]

Departments of Virology[#] and Medicinal Chemistry*
Sterling Research Group
Rensselaer, NY 12144

INTRODUCTION

Chemotherapeutic agents such as acyclovir and zidovudine have been extremely useful in treating patients with herpes simplex virus infections or AIDS, respectively. However, there is another scientific use for drugs or compounds which is to use them as molecular tools in the research laboratory in order to dissect the replication of viruses and to discover new facts about viruses. For example, alpha-amanitin, an inhibitor of the cellular DNA-dependent RNA polymerase II (Pol II), is used to define whether a particular virus RNA species is synthesized with the involvement of Pol II or solely by viral polymerases. With herpesviruses, cycloheximide, an inhibitor of protein synthesis, is used to define the "immediate early" class of viral genes. If a viral RNA is synthesized in the presence of cycloheximide, then the synthesis of that viral RNA is not dependent on the synthesis of any viral protein and that viral gene can be classified as immediate early. Arabinosyl cytosine (AraC) is an anticancer/antiherpesvirus drug which inhibits DNA synthesis in cell culture systems but allows RNA and protein synthesis to proceed. One can ask whether the replication of an RNA virus is dependent on cellular DNA synthesis by determining if virus replication is sensitive to AraC. These examples demonstrate how chemotherapeutic agents can be used as molecular tools in the research laboratory.

The purpose of this manuscript is to discuss several new biological concepts that we have learned by using antipicornaviral compounds as molecular tools. These compounds bind in a hydrophobic pocket in viral protein l (VP1) of the rhinovirus capsid as demonstrated by X-ray crystallography (Smith, et al., 1986; Badger et al., 1988). Amino acid substitutions in the rhinovirus capsid can prevent compound binding and reduce the sensitivity of the virus to the inhibitory effects of the compounds (Heinz, et al., 1989). Thus, the antiviral activity of this class of antipicornaviral compounds is a sensitive probe for the amino acids of the

Innovations in Antiviral Development and the Detection of Virus Infection
Edited by T. Block *et al.*, Plenum Press, New York, 1992

hydrophobic pocket of VP1 and the conformational changes induced upon drug binding. In this manuscript, we will discuss how results from studies using these compounds have demonstrated that: 1) the canyon hypothesis (i.e., that a canyon on the surface of the picornavirus virion is the binding site for the cellular receptor) is supported by the observation that the conformational changes induced in HRV-14 by compound binding are sufficient to inhibit viral adsorption to cellular receptors; 2) the structures of most major binding group HRVs may be similar to each other, but different from most minor binding group HRVs; 3) human rhinoviruses uncoat faster and in a different cellular compartment than human polioviruses, 4) the binding affinity of several antiviral compounds to HRV-14 correlates with antiviral potency, and 5) in contrast to other enteroviruses, the prototypic strain (Nancy) of coxsackievirus B3 (CB3) is not representative of clinical isolates of CB3.

RESULTS AND DISCUSSION

Supporting evidence for the Canyon Hypothesis

The canyon hypothesis states that the host cell receptor for picornaviruses binds in a 25 Å deep canyon on the virion surface. The canyon is too narrow for antibodies to enter, but the cellular receptor (ICAM-1 for the major binding group of HRVs) is hypothesized to be narrow enough to bind to the canyon during viral adsorption to host cells. The capsid-binding compounds were shown to induce significant conformational changes in the bottom of the canyon of HRV-14(Badger, et al., 1988). The canyon hypothesis predicts that the capsid-binding compounds would interfere with HRV-14 adsorption to ICAM-1. In studies performed with eight compounds and HRV-14, the concentration of compound required to inhibit the binding of radiolabelled virus correlated with the antiviral activity as measured in plaque reduction assays (r^2=0.81)(Pevear, et al., 1989). These experimental results, along with results with site-directed mutants in the canyon of HRV-14 (Colonno, et al., 1988), support the canyon hypothesis.

Predictions of the virion structure of major and minor HRVs

The capsid-binding compounds have been shown to induce significant conformational changes in the virion upon binding to HRV-14 (a major group HRV; Badger et al., 1988) but to not induce any significant conformational changes in HRV-1A (a minor group HRV; Kim et al., 1989). Mechanism of action studies have demonstrated that the compounds inhibit the adsorption of HRV-14 but not HRV-1A (Pevear, et al., 1989; Kim et al., 1989). We have extended these studies to three additional major group HRVs and two additional minor group HRVs (Figure 1;Structures of compounds in Figure 2). The results clearly demonstrate that this antiviral compound inhibited the adsorption of the 4 major HRVs but, in contrast,

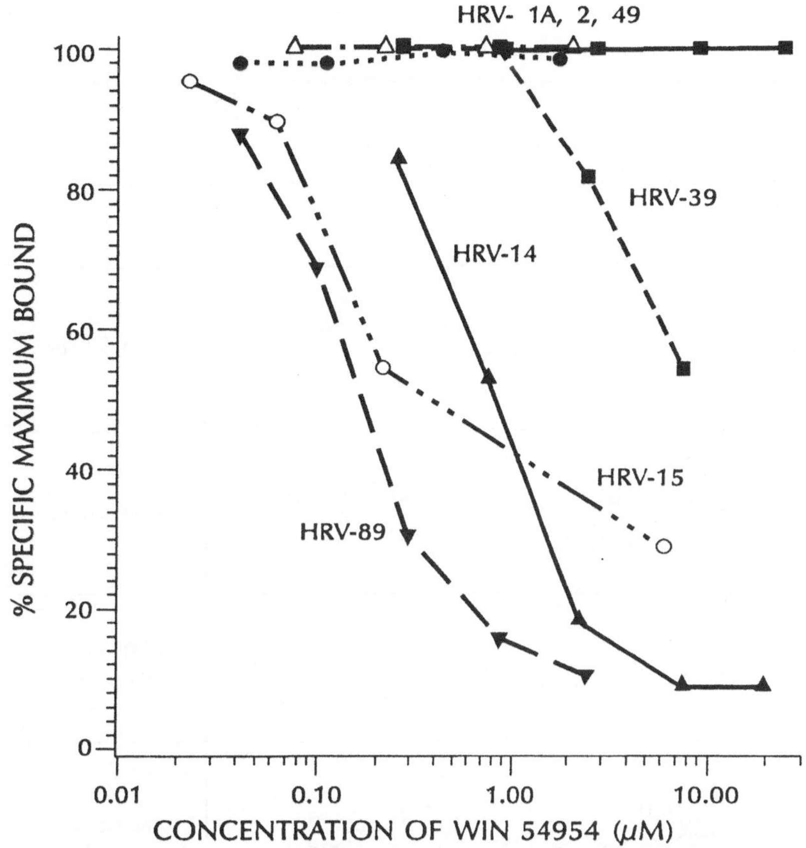

Figure 1. Dose-response curves of the adsorption of different HRV serotypes to HeLa cells. HRV serotypes were labeled with [^{35}S] methionine and purified by gradient centrifugation. Win 54954 was solubilized and serially diluted in DMSO, and preincubated with radioactive HRV for 1 hour at 20°C. Intact HeLa cells (5 x 10^5 per reaction) were then added to the HRV-drug mixture and incubated for 2 hours at 20°C. Cells were pelleted by centrifugation and the % of radioactivity bound to cells and in the supernatant were determined. The % binding was normalized to the maximum binding for each serotype, and this % specific maximum bound was plotted versus the concentration of WIN 54954.

Major binding group HRVs (which bind to the cellular ICAM-1) are HRV-89 ▼— —▼, HRV-14 ▲———▲, HRV-15 ○——— ○ and HRV-39 ■- — — -■, Minor binding group HRVs (which bind to cellular receptors other than ICAM-1) are HRV-1A ■———■, HRV-2 ▲— · —▲ and HRV-49 ● · · · · · ·●.

STRUCTURE	WIN
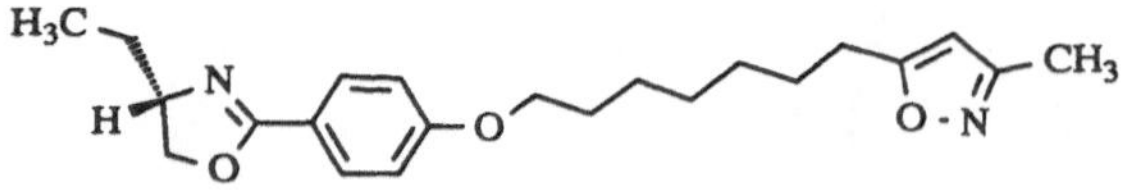	WIN 56590
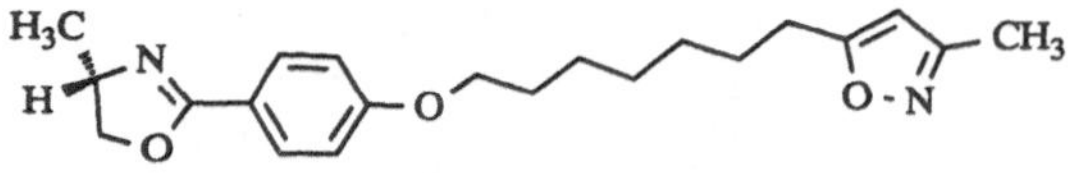	WIN 52084
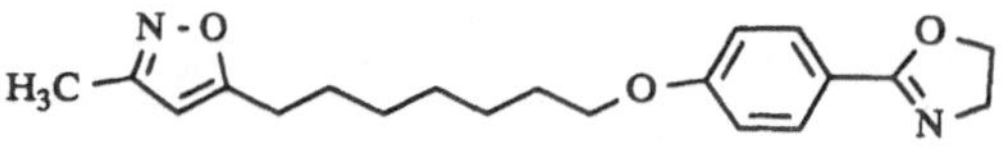	DISOXARIL (WIN 51711)
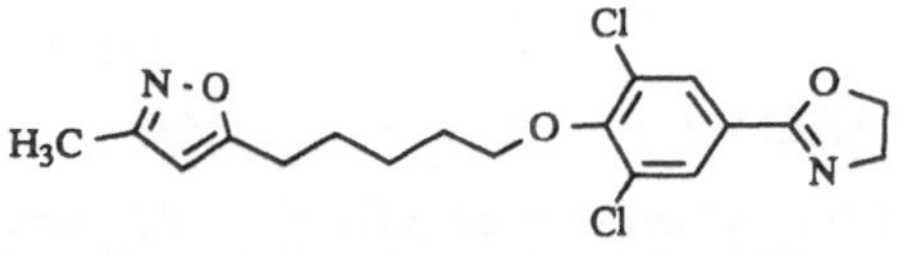	WIN 54954

Figure 2 . The chemical structures of WIN 56590, WIN 52084, disoxaril, and WIN 54954.

Table 1. Kinetics of Rhinovirus and Poliovirus Uncoating

Virus[a]	% Virus Yield[b]	Time of Addition (hours post infection)
Rhinovirus	1	0.0
	60	0.25
	100	0.75
	100	4.0
Poliovirus	1	0.0
	1	0.25
	30	0.75
	100	4.0

[a]Rhinovirus=HRV-2 ; Poliovirus=PV-2

[b]The % of control virus yield was determined by infecting Hela cells with HRV-2 or PV-2 and adding disoxaril at various times after infection. The virus yield was determined by plaque assay.

the compound did not inhibit the adsorption of the 3 minor HRVs. These results suggest that the capsid-binding compounds will induce significant conformational changes in all major HRVs but no significant conformational changes in the minor HRVs. In addition, these results suggest that the virion structures of the major group HRVs will be similar to each other, but significantly different from the minor group HRVs. These hypotheses will be the focus of future crystallographic and mechanism of action studies.

Uncoating of Rhinoviruses and Polioviruses

In order to examine the kinetics of viral uncoating, time of addition studies were performed with disoxaril (WIN 51711; Figure 2) and HRV-2 and poliovirus type 2 (PV-2), two viruses for which inhibition of uncoating is the only known mechanism of action (Fox, et al., 1986). If disoxaril was present at the time of infection, then the replication of both HRV-2 and PV-2 was inhibited. However, if disoxaril was added to infected cells at various times after infection (Table 1), the inhibitory effect was lost at an earlier time for HRV-2 than for PV-2. For example, HRV-2 replicated to approximately 60% of the control yield if disoxaril was added at 15 minutes after infection. In contrast, PV-2 replicated to only 1% of the control yield under similar conditions. These results demonstrate that HRV-2 uncoated at a faster rate than PV-2 and were confirmed using neutral red labelled HRV-2 and PV-2.

These results suggest that rhinoviruses may uncoat in a different manner or cellular compartment than polioviruses. These differences may represent significant factors in addition to viral receptors with respect to viral tropism, ie. the organ systems and cell types in an organism which are productive sites of replication and tissue injury. For example, the restricted replication of human rhinoviruses in nasal epithelium may be a consequence of a specialized cell type in the nasal epithelium which possesses an appropriate cellular compartment needed for rhinovirus uncoating and subsequent productive infection.

Binding Parameters of Radioactive Compounds to HRV-14

In an effort to determine whether the binding affinity of compounds for HRV-14 correlated with antiviral potency, radioactive compounds were synthesized and examined for their binding affinity to purified HRV-14 by Scatchard analysis (Fox, et al., 1991). Binding of compounds to virions was specific and saturable. The results are summarized in Figure 3. The correlation between the MIC and the binding affinity was very high for the 4 compounds (r=0.997). In addition, no detectable binding of WIN 54954 could be observed with a drug resistant mutant of HRV-14 in which the valine at position 188 of viral protein of wild type virus has been substituted with leucine. These results demonstrate that there is a direct relationship between antiviral activity and binding affinity. Compounds with more potent antiviral activity have a higher binding affinity to HRV-14. This conclusion is true for this limited series of compounds but may not be true for other chemical classes of compounds.

While there was a direct relationship between affinity and antiviral activity, no relationship was observed between binding affinity and either the independently determined rate of association (Kon) or the rate of dissociation (Koff). For example, the compound with the fastest Kon (disoxaril) did not have the highest binding affinity (Figure 3). In addition, the compound with the lowest binding affinity (WIN 54954) did not have the fastest Koff. These anomolies were resolved through the relationship of these binding parameters (KD=Koff/Kon). Schematically, this relationship is shown in Figure 3 where the distance between the vertical dashed lines indicates the magnitude of the binding affinity. Thus, one can understand that the slow Koff rate of WIN 52084 is the major determinant of the high binding affinity, and how the compound with the intermediate binding affinity (disoxaril) actually has the fastest rate of association as well as the fastest rate of dissociation.

The direct relationship between binding affinity and antiviral activity is an important concept for rational design of more active compounds. An analysis of the spatial aspects of compound binding indicates that the synthesis of compounds which fill the space more completely in the hydrophobic pocket in the virus may result in compounds with higher binding affinities and higher antiviral potency.

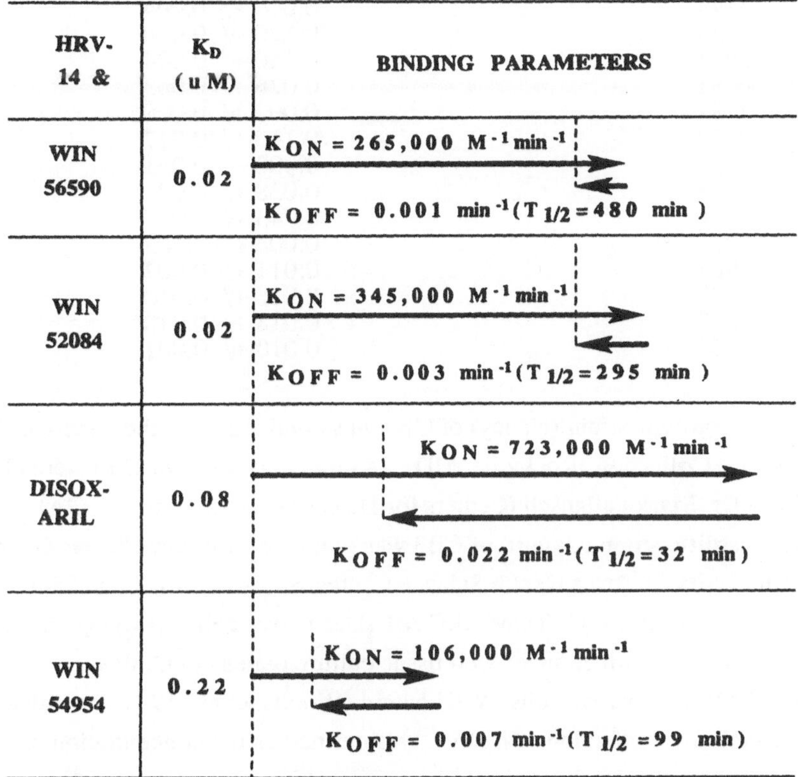

HRV-
14 &
KD
(u M)
BINDING PARAMETERS
WIN
56590
0.02
KON = 265,000 M-1 min-1
KOFF = 0.001 min-1 (T 1/2 = 480 min)
WIN
52084
0.02
KON = 345,000 M-1 min-1
KOFF = 0.003 min-1 (T 1/2 = 295 min)
DISOX-
ARIL
0.08
KON = 723,000 M-1 min-1
KOFF = 0.022 min-1 (T 1/2 = 32 min)
WIN
54954
0.22
KON = 106,000 M-1 min-1
KOFF = 0.007 min-1 (T 1/2 = 99 min)

Figure 3. The binding parameters for HRV-14 and 4 antiviral compounds are shown schematically by the length of the arrows. A fast K_{on} (disoxaril) is represented by a longer arrow than a slow K_{on} (WIN 54954). A short half life (disoxaril) is represented by a longer arrow than a long half life (WIN 56590). The distance between the vertical 2 dotted lines represents the binding affinity (K_D). A high binding affinity (WIN 56590 or WIN 52084) has a greater distance

Table 2. *In Vitro* Activity of WIN 54954 Against Coxsackievirus B3

Strain or Isolate	Mean MIC +/- Std. Dev. (μg/ml)
Nancy (prototype)	> 3.1 (not active)
Gauntt	0.075 +/- 0.056
3174	0.014 +/- 0.004
5835	0.120 +/- 0.060
5965	0.140 +/- 0.004
6781	0.026 +/- 0.003
7782	0.009 +/- 0.0005
9338	0.110 +/- 0.017
9834	0.400 +/- 0.031
1068	0.038 +/- 0.007
3378	0.043 +/- 0.050
2225	0.006 +/- 0.003
6907	0.011 +/- 0.002
6408	0.016 +/- 0.010
5553	0.018 +/- 0.012
1964	0.010 +/- 0.001

[a]The prototype strain (Nancy) of CB3 was obtained from the American Type Culture Collection (Rockville, MD), 14 clinical isolates of CB3 were obtained from Dr. Mark Pallansch (Centers for Disease Control, Atlanta, GA), and a myocarditic strain (Gauntt) of CB3 was obtained from Dr. Charles Gauntt (University of Texas Health Science Center, San Antonio, TX). Virus stocks were prepared from infected LLC-MK_2Derivative cells. A cytopathic effect assay was performed in 96-well tissue culture plates containing LLC-MK_2Derivative cells. WIN 54954 (Woods, et.al., 1989) was added to the cells after infection. The MIC was defined as the concentration of compound which inhibited the virus cytopathic effect by 50%. The mean +/- the standard deviation is shown.

Coxsackievirus B3

The results in Table 2 show that the prototypic Nancy strain of CB3 was not sensitive to WIN 54954. In sharp contrast to this result, the 14 clinical isolates of CB3, as well as the myocarditic strain of CB3, were all sensitive to WIN 54954. The MICs for the sensitive strains of CB3 ranged from 0.006 to 0.400 ug/ml, with an average (+/- the standard deviation) MIC of 0.069 (+/- 0.100) μg/ml.

These results demonstrate that the sensitivity of the Nancy strain of CB3 to WIN 54954 is clearly different from the 14 clinical isolates and the myocarditic Gauntt strain of CB3. In other studies not shown here (manuscript in preparation), WIN 54954 was tested in a similar manner against prototypic strains and 204 clinical isolates of the 15 most commonly isolated enteroviruses (Strikas, et.al., 1986). With all other enteroviruses, the MIC for the prototypic strain was generally consistent with the clinical isolates. Furthermore, all 204 clinical isolates were sensitive to WIN 54954. The data suggest that the drug binding pocket in VP1 of CB3(Nancy) differs significantly from the clinical isolates of CB3.

CONCLUSIONS

The results in this manuscript have shown that compounds can be extremely useful molecular tools in the research laboratory. Compounds that are known to bind in a hydrophobic pocket in the picornaviral capsid can be used to probe the amino acids lining that pocket. Using thse compounds as molecular probes, one can uncover supporting evidence for the canyon hypothesis, predict that structures of other rhinovirus serotypes, find out that rhinoviruses uncoat faster than polioviruses, observe that the binding affinity of these compounds correlates with antiviral activity, and reveal that a prototypic strain of one enterovirus is not representative of clinical isolates .

REFERENCES

Badger, J., Minor I., Kremer, M.J., Oliveira, M.A., Smith, T.J., Griffith, J.P., Geurin, D.M.A., Krishnaswamy, S., Luo, M. Rossmann, M.G., McKinlay, M.A., Diana, G.D., F.J. Dutko, F.J., M. Fancher, M., Reuckert, R.R., and Heinz, B., Structural Analysis of a Series of Antiviral Agents Complexed With Human Rhinovirus 14, Proc. Natl. Acad. Sci. USA 85:3304 (1988).

Colonno, R.J., Condra, J.H., Mitutani, S., Callahan, P.L., Davies, M.-E., and Murcko, M.A., Evidence for the Direct Involvement of the Rhinovirus Canyon in Receptor Binding, Proc. Natl. Acad. Sci. USA 85:5449 (1988).

Fox, M.P., Otto, M.J., and McKinlay, M.A., Prevention of Rhinovirus and Poliovirus Uncoating by WIN 51711, A New Antiviral Drug, Antimicrob. Ag. and Chemo. 30:110 (1986).

Fox, M.P., McKinlay, M.A., Diana, G.D., and Dutko, F.J., Binding Affinities of Structurally Related Human Rhinovirus Capsid-Binding Compounds are Related to Their Activities Against Human Rhinovirus Type 14, Antimicrob. Ag. and Chemo., 35 (6), In Press, June, 1991.

Heinz, B.A., Reuckert, R.R., Shepard, D.A., Dutko, F.J., McKinlay, M.A., Fancher, M., Rossmann, M.G., Badger, J., and Smith, T.J., Genetic and Molecular Analyses of Spontaneous Mutants of Human Rhinovirus 14 that are Resistant to an Antiviral Compound, J. Virol. 63:2476 (1989).

Kim, S., Smith, T.J., Chapman, M.S., Rossmann, M.G., Pevear, D.C., Dutko, F.J., Felock, P.J., Diana, G.D., and McKinlay, M.A., Crystal Structure of Human Rhinovirus Serotype 1A (HRV1A), J. Mol. Biol.210:91 (1989).

Pevear, D.C., Fancher, M.J., Felock, P.J., Rossmann, M.G., Miller, M.S., Diana, G., Treasurywala, A.M., McKinaly, M.A., and Dutko, F.J., Conformational Change in the Floor of the Human Rhinovirus Canyon Blocks Adsorption to HeLa Cell Receptors, J. Virol. 63:2002 (1989).

Smith, T.J., Kremer, M.J., Luo, M., Vriend, G., Arnold, E., Kamer, G., Rossmann, M.G., McKinlay, M.A., Diana, G.D., and Otto, M.J., The Site of Attachment in Human Rhinovirus 14 for Agents that Inhibit Uncoating, Science 233:1286 (1986).

Strikas, R.A., Anderson, L.J., and Parker, R.A., Temporal and Geographic Patterns of Isolates of Nonpolio Enterovirus in the United States, 1970 - 1983, J. Inf. Dis. 153:346 (1986).

Woods, M.G., Diana, G.D., Rogge, M.C., Otto, M.J., Dutko, F.J., and McKinlay, M.A., In Vitro and In Vivo Activities of WIN 54954, A New Broad-Spectrum Antipicornavirus Drug, Antimicrob. Ag. and Chemo. 33:2069 (1989).

GENETICALLY ENGINEERED BACTERIA TO IDENTIFY AND PRODUCE ANTI-VIRAL AGENTS

Robert H. Grafstrom, Katherine Zachariasewycz,
Richard A. Brigandi, and Timothy M. Block

Department of Microbiology and Immunology
Jefferson Medical College
Thomas Jefferson University
Philadelphia, PA 19107

SUMMARY

We have prepared a strain of *Escherichia coli* that expresses both the HIV protease and a Tet protein which has been modified to contain the HIV protease recognition sequence. When the protease is expressed, the bacteria will not grow in the presence of tetracycline. However, when the protease is inhibited the bacteria can grow in tetracycline containing media (Block and Grafstrom 1990). We have selected spontaneously arising Tet resistant mutants and have screened them for those that could be producing an inhibitor of HIV protease. The problems in the construction of this strain and the characterization of the various Tet^r mutants are discussed.

INTRODUCTION

Since the isolation of antibiotics, interest in bacteria both as the source of biologically important molecules and as a solution to environmental problems has been a leading impetus for the study of microorganisms. In addition to the hundreds of antibiotics, the anticancer agent, adriamycin (Middleman et al., 1971), the antiviral agent adenosine arabinoside (Sidwell et al., 1968), and the immune modulator, cyclosporin A (Borel et al., 1977), have all been isolated from microorganisms. Moreover, a bacterial strain has been developed to screen chemical compounds for carcinogenic potential using a bacterial mutagenesis assay (Ames et al., 1975). Recently, bacteria with other special properties such as the ability to dissolve oil spills or to prevent frost formation on crops have been important additions to our repertoire of useful bacteria. With the advent of recombinant DNA techniques it was clear that recombinant bacteria could also be turned into factories to produce new and exciting drugs that could not otherwise be readily synthesized in the laboratory. These now include such well known proteins as insulin, growth hormone, clotting factor, etc. The use of recombinant bacteria has also been invaluable in the study of clinically important bacteria and viruses. These efforts have increased the safety of studying very hazardous organisms and have produced high levels of proteins necessary for both

Innovations in Antiviral Development and the Detection of Virus Infection
Edited by T. Block *et al.*, Plenum Press, New York, 1992

biochemical studies and the production of safe vaccines. Recombinant bacteria have been useful in the identification of mammalian DNA binding proteins (Kadonaga et al., 1987), the characterization of trypsin structural gene mutations (Evnin et al., 1990), in the isolation of mutant HIV protease (Baum et al., 1990a; Baum et al, 1990b), in the production of diverse antibody libraries (Huse et al., 1989), and in the study of receptor-ligand interaction (Cwirla et al., 1990). Based upon these studies, we have proposed that a new era in human use of bacteria could be signaled by "designer bacteria" to identify (and perhaps even generate) antiviral agents or for that matter any "anti-clinically-important-protein."

We propose to use genetically engineered microorganisms to achieve two related objectives. *First*, microorganisms will be used to screen for inhibitors of clinically important viruses (or any clinically important proteins such as onco-proteins); and, second, by applying the power of bacterial genetics, mutant bacteria will be selected that could themselves produce inhibitors of these biologically important proteins. Both objectives have been demonstrated by us using the bacteria *E. coli*. The application of this concept first requires that the growth of bacteria be made dependent on the inhibition of the desired protein (i.e. conditional lethal). Second, the growth in selective media allows for selection of mutant bacteria that have somehow circumvented the toxic effects of the cloned protein. This process can only be applied if the biochemical activity of the target protein is understood and if the protein can be expressed and is operational in bacteria.

There are numerous reasons for developing a bacterial screen to search for antiviral agents. Safety is a major impetus since many of the assays for antiviral agents use the growth of live virus which can be dangerous. Secondly, no convenient cell culture system exists for some important viruses and one is left with the use of animal models. Many of the animal and tissue culture assays are very complex and demand a high degree of technical sophistication. They are also time consuming and costly, and as such are impractical as an initial screen for antiviral agents. Thirdly, although the biochemical assays have the advantages of speed and specificity, they are not as useful for the screening of crude chemical mixtures which could otherwise non-specifically inhibit these enzymologically based assays. Finally, many pharmaceutical companies already have in place bacterial screens for antibiotics and the use of the "designer bacteria" described here to screen for antivirals would dovetail nicely with this approach. However, it should be pointed out that the use of a bacterial screen is only meant as an initial means to identify potential antiviral compounds and not to supplant the use of other more refined assays.

RESULTS AND DISCUSSION

The Assay

To demonstrate our concept we have chosen the Human Immunodeficiency virus (HIV) protease since HIV is the causative agent of AIDS and as yet there is only one approved drug for its treatment, AZT, which inhibits the replication of the virus by blocking the reverse transcriptase (DeClercq 1990; Mitsuya et al.,). The HIV protease has been shown to be essential for viral growth, since mutants that lack the protease do not produce infectious virus (Debouk et al., 1987; Kohl et al., 1988; Legrice et al., 1988). Consequently, the protease has been the object of an intense research effort to identify a new anti-HIV agent (DeClercq 1990; Mitsuya et al.;

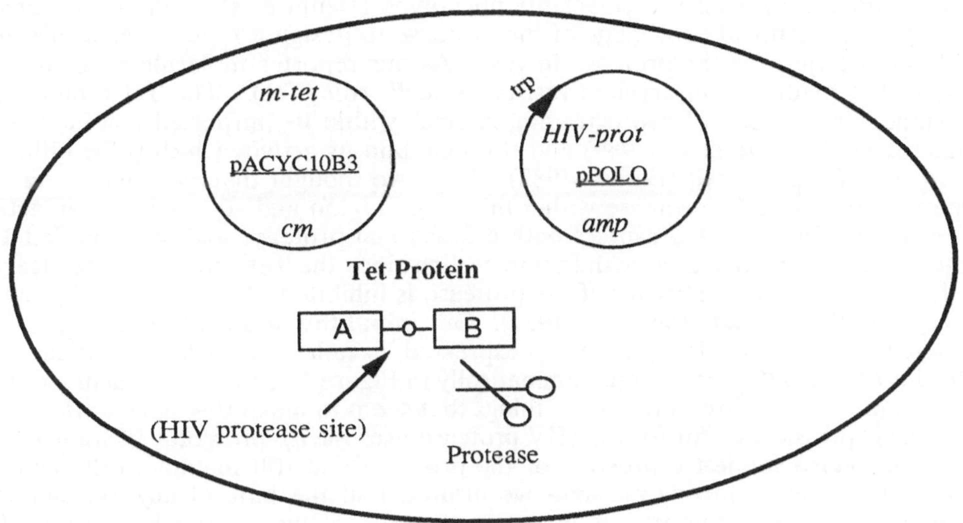

In the **absence** of an inhibitor of the HIV protease, the bacteria are sensitive to tetracycline

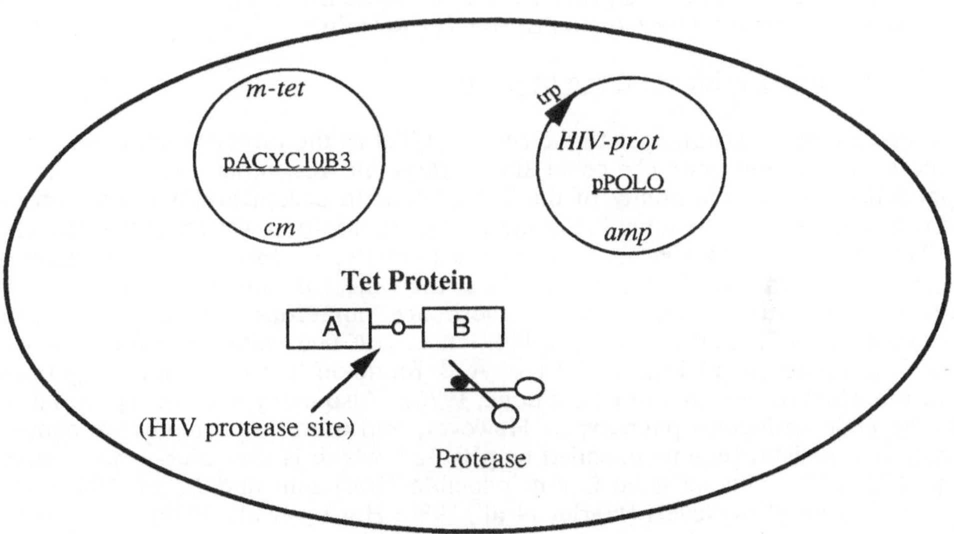

In the **presence** of an inhibitor of the HIV protease, the bacteria are resistant to tetracycline

FIGURE 1

Moore et al., 1989). Since a functional protease could be expressed in bacteria, we thought it was an ideal candidate for the demonstration of our concept. However, although the over expression of the protease has been demonstrated to be lethal to *E. coli* (and this in itself is a selectable phenotype (Baum et al., 1990b), we wanted to use the biochemical knowledge of the protease to design a reporter molecule that would be a target for the protease in vivo. As our reporter molecule we chose the protein that mediates tetracycline resistance in *E. coli* (Tet). The Tet protein can withstand the addition of two extra amino acids within its purported non-functional hinge region (Curiale et al., 1984) and still maintain its activity which is the efflux of tetracycline from the cell (Barany 1985). Thus, we thought that we could insert the target site for the HIV protease within this hinge region and still maintain an active Tet protein. Bacteria that express both a functional protease and the modified Tet protein will be sensitive to growth in tetracycline since the Tet protein will be cleaved by the HIV protease. However, if the protease is inhibited, the bacteria will grow in tetracycline (Block and Grafstrom, 1990). Our initial test strain which is sensitive to tetracycline, when the HIV protease is expressed, is called TJU525. The design and behavior of TJU525 is shown diagrammatically in Figure 1. In the construction of this test system we have discovered many things that seem to make this assay work. First, since the expression vector for the HIV protease uses the *trp* promotor (Korant 1990), we could induce modest expression of the protease and still maintain cell viability. Second, by using multicopy plasmids we insured that the bulk of any Tet^r mutants isolated would be chromosomal in nature since it is most unlikely to select for mutations that would have occurred in each copy of the plasmid (see below). Finally, we were also aware that other viral proteases which have been expressed in *E. coli* and are functional would be logical targets for this type of assay (Ivanoff et al., 1986). Of course, for each individual protease a new protease recognition site would have to be inserted into the hinge region of the Tet protein.

The Tetracycline Resistance Gene of *E.coli*

There are many advantages to the choice of Tet as the target protein. At the start of these experiments our choice of the tetracycline resistance protein as a target depended solely on the ability of the Tet protein to accept additional amino acids and still maintain its phenotype (Barany 1985). In addition, we knew that the use of the Tet protein would give us a selectable phenotype. Moreover, during the last decade the regulation of *tet* expression was the object of intensive biochemical and genetic investigation. For instance, there are numerous classes of tetracycline resistance genes (Rubin and Levy 1990). In *E. coli* five classes are found of which class C is found in pBR322 and Class A is found in TN10. Some *tet* genes are inducible while others are not (Yang et al., 1976). Also many bacterial species display a tetracycline resistance phenotype. However, and more importantly, we chose the tetracycline resistance gene encoded by pBR322, which is the same as that encoded by pACYC184. This *tet* gene is not inducible (Backman and Boyer, 1983) and is expressed at very low levels (Harley et al., 1988; Harley et al., 1990). It is probably this low expression of the *tet* gene coupled with the low level of expression of the protease that makes our assay successful. Too much expression of either the protease or the target protein would not have resulted in a functional assay. In addition, the use of *tet* made it possible to select threshold levels of tetracycline to fine tune the bacteriological assay. Finally, and most importantly, of the five tetracycline resistant classes isolated from *E. coli* each are plasmid encoded and each mediate resistance by the active efflux of tetracycline from the cell (Marshall et al., 1986; McMurry et al., 1980). Thus, chromosomal mutations that result in tetracycline resistance in *E. coli* have not yet been identified suggesting that such mutations are lethal. In addition, it is possible that by the use of a drug resistance gene that catalyzes the efflux of the antibiotic from the cell it is possible that we might identify a mutant which will secrete an HIV protease inhibitor using this same efflux pathway. If our knowledge of bacteria were even more extensive we are sure that equally suitable, non-Tet markers could be found but we point out these salient features so that others will consider some of the problems we faced (and serendipitously avoided) in the development of this prototypic bacteria.

The Permeability of the Bacterial Membrane

In discussions with others, a major objection to our concept has been the permeability of the bacterial membrane. Initially, we thought that this problem could be overcome by the use of organisms other than *E. coli*, for instance gram positive organisms or even yeast. In fact, one laboratory has been studying the human topisomerase in yeast and looking at the resistance to the drug camptothecin (Bjornsti et al., 1989). However, there is a wealth of evidence for the many types of chemicals that do cross the bacterial membrane. As a direct result of the Ames test, many compounds have been identified both for their ability to induce bacterial mutagenesis and to cause cancer in animals. An initial screen of over 300 chemicals by Ames and coworkers (McCann et al., 1975) shows numerous classes of chemicals are mutagenic in bacteria and presumably penetrate the membrane. More recently, comparison of the mutagenic and carcinogenic potential of 222 chemicals was published (Ashby and Tennant, 1988). A quick review of the listed compounds, which are mutagenic in bacteria, reinforces the concept that bacteria have the capacity to internalize a broad spectrum of chemical classes, many of which are also taken up by mammalian cells. Parenthetically, it would be interesting to know if the compounds that are carcinogenic in animal studies by not mutagenic in bacteria are not taken up by bacteria or represent a more basic underlying mechanism of mutagenesis that is not shared by the bacteria. Thus, although we feel that permeability is not a major problem in this assay (and the data support this), it is possible that some anti-viral compounds could be missed. However, by the use of a battery of different genera of test or bacteria then this problem could probably be reduced.

The Isolation of Tetracycline Resistant Mutants

The construction of TJU525 and its phenotype have all been documented in detail (Block and Grafstrom, 1990). Although this strain is valuable as a screen for the identification of inhibitors of the HIV protease, we have limited our analysis to the verification of the assay with the use of Pepstatin A, a known inhibitor of the HIV protease (Hansen et al., 1988; Krausslich et al., 1988; Seelmeir et al., 1988). Instead, our focus has been on the isolation and characterization of the Tet resistant mutants that we have selected. As stated above TJU525 does not grow in tetracycline containing media since its Tet protein is cleaved by the HIV protease. This allows for the genetic selection of rare mutants that have regained the ability to grow in tetracycline. Among the Tet resistant mutants would be ones that have reverted the HIV protease site within our modified Tet protein. Other Tet resistant mutants could have inactivated the HIV protease. However, as unlikely as it might seem, it is possible that among these Tet^r mutants are ones that themselves are producing inhibitors of the HIV protease. In this latter category are *E. coli* mutants that might degrade the HIV protease, block its expression (e.g. over producers of tryptophan), or ones that make a protein that complexes with the protease and form an inactive complex. In addition, it could be possible to generate Tet^r mutants that would make a low molecular weight inhibitor of the protease. Both the Lac and Trp repressors are just two well known examples of a strategy whereby *E. coli* uses low molecular weight effector molecules to greatly alter protein activity. It is this last class of mutants that could provide an innovative class of antiviral agents or furnish a model compound for the design of an effective antiviral agent. The difficulty, as we will discuss below, is

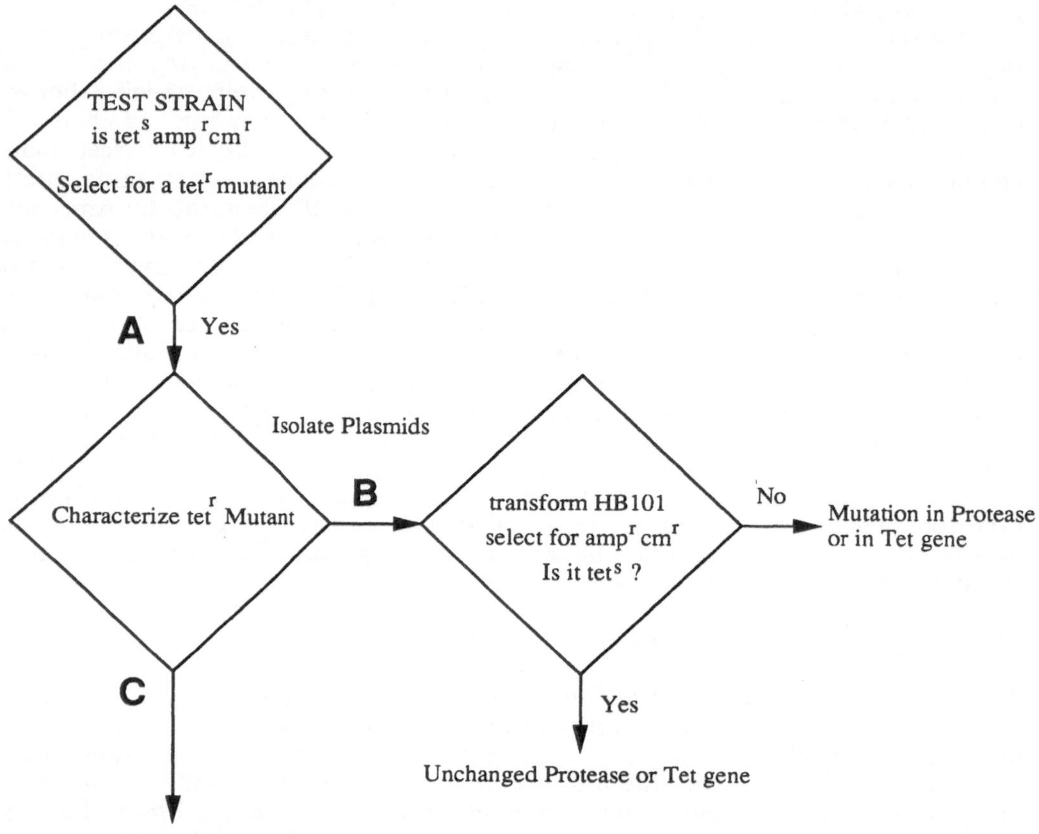

CO-CULTIVATION ASSAY

Check for Production of Inhibitor of Viral Protease

Mix Tet resistant (white) mutants with Tet sensitive (blue) testor strain and incubate on plates containing ampicillin, chloramphenicol, and tetracycline

Figure 2. Flow diagram for the characterization of teracycline resistant mutants

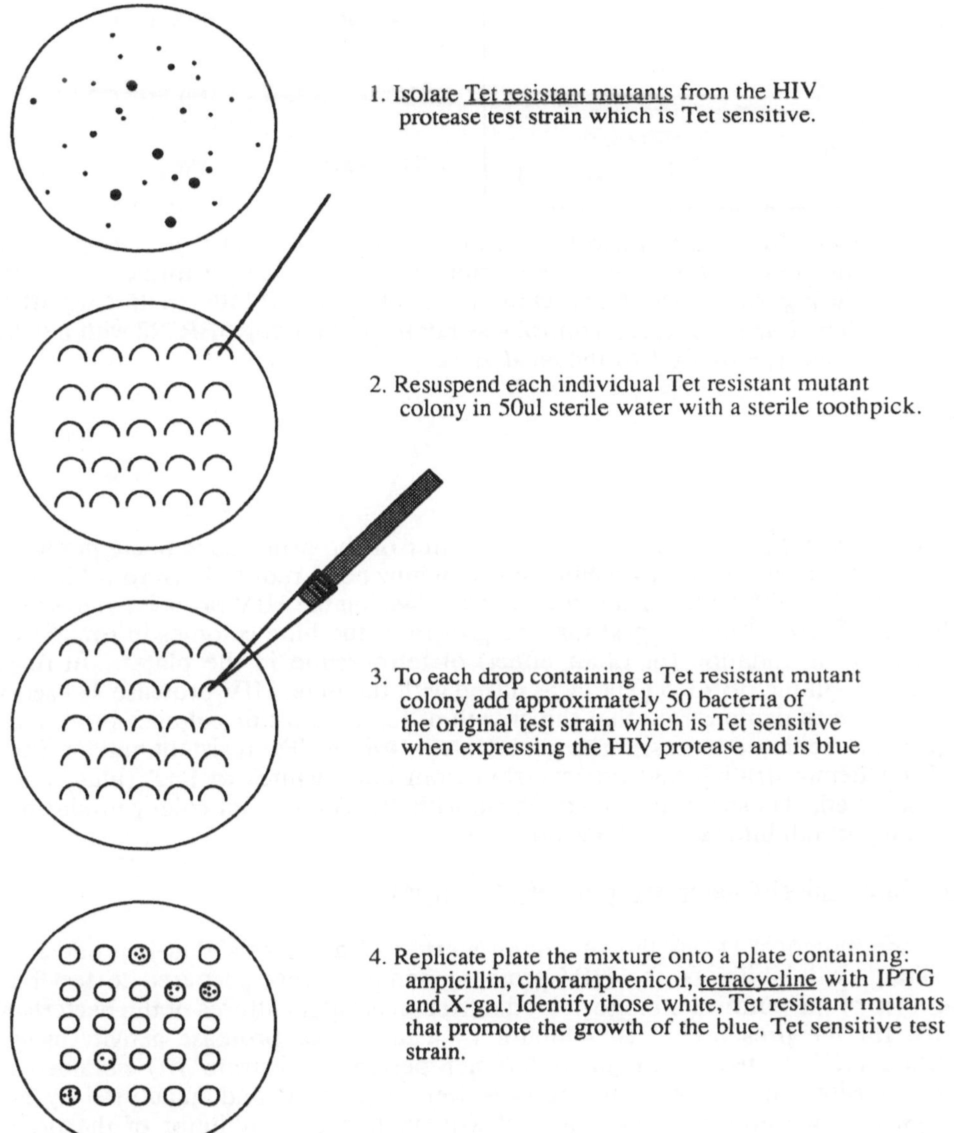

1. Isolate Tet resistant mutants from the HIV protease test strain which is Tet sensitive.
2. Resuspend each individual Tet resistant mutant colony in 50ul sterile water with a sterile toothpick.
3. To each drop containing a Tet resistant mutant colony add approximately 50 bacteria of the original test strain which is Tet sensitive when expressing the HIV protease and is blue
4. Replicate plate the mixture onto a plate containing: ampicillin, chloramphenicol, tetracycline with IPTG and X-gal. Identify those white, Tet resistant mutants that promote the growth of the blue, Tet sensitive test strain.

Figure 3. Co-cultivation assay for the identification of mutants that could be synthesizing putative inhibitors of the HIV protease

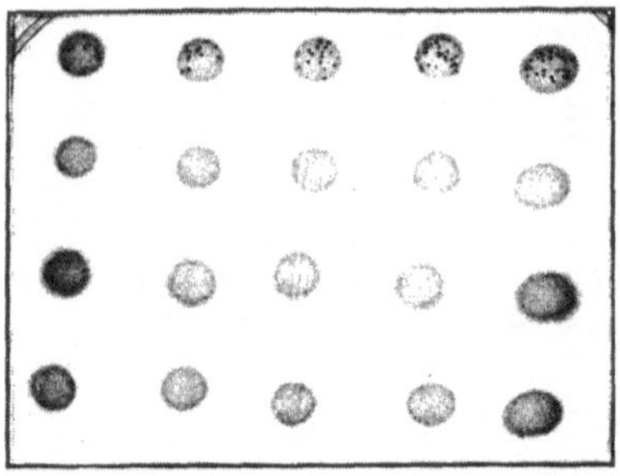

	STRAIN	
	Tet resistant (white)	Tet sensitive (blue)
1.	Mutant #11	HIV testor strain
2.	Mutant #11	Control
3.	Control	HIV testor strain
4.	Mutant #11	None

Figure 4. Co-cultivation assay to test for mutants which could be producing inhibitors of the HIV protease. The co-cultivation assay was performed as described in Figure 3. The Lac-, white, tet^r control was JM109 containing pBR325. The Lac+, blue, tet^s control was HB101 containing pBR322 with a deletion from the *Eco*RV to the *Ava*I sites.

consistent with the idea that either an inhibitor of the protease is being produced or that the effectiveness of tetracycline has somehow been reduced. In row #2, mixture of mutant #11 with a blue, Tet sensitive strain lacking the HIV protease yields no blue colonies. These results suggest that the growth of the blue colonies in row #1 is not due to the degradation (or other effect) of tetracycline in the plates. In row #3, JM109 containing pBR325 (Tet^r) was mixed with the blue, HIV protease Tet sensitive testor strain to demonstrate that the growth of the blue testor colonies is not a result of plasmid mediated tet resistance gene. In row #4, the Tet resistant mutant is plated by itself demonstrating that no reversion from Lac$^-$ (white) to Lac$^+$ (blue) colonies had occurred. These results are consistent with the Tet^r mutant colony producing and secreting an inhibitor of the HIV protease.

The Biochemical Characterization of Mutant #11

Since mutant #11 has the genetic properties of a bacteria that is producing and secreting an inhibitor of the HIV protease, samples were prepared to test for the presence of the putative inhibitor. Media from overnight cultures of the bacteria were tested for the presence of an inhibitor by assaying for protease activity using the purified HIV protease and an undecamer peptide (substrate III) obtained from Bachem Bioscience. Reaction products were separated and quantified by HPLC analysis. The standard assay with and without a known inhibitor of the protease, Pepstatin A (Hansen et al., 1988; Lraiss; ocj et al., 1988) is shown in Figure 5. In the absence of Pepstatin A, there is a 60% cleavage of substrate III (s) into the carboxy (c) and amino (n) terminal peptides. In the presence of Pepstain A, this reaction is inhibited by 90%. The same reaction performed in the presence of media isolated from mutant #11 or from its parent (Wildtype) is shown in Figure 6. In each case, a mechanism by which to distinguish between each of these different classes of mutation.

It should be pointed out that in the selection of mutants we have made use of the recent experiments by Cairns (Cairns et al., 1988) and Hall (Hall, 1990). In their experiments, the mutation frequency of *E. coli* is enhanced when the mutants are isolated in the presence of a selective agent. In our case this is tetracycline. Although there is much controversy surrounding the interpretation of the results of Cairns and Hall, we think it is important to note that our Tet^r mutants were isolated using the information they have published and could be interpreted as extending their conclusions to include a drug resistant gene since their experiments were done with auxotrophic mutants. Whether or not the Tet^r mutants reported here are actually the result of a "Cairnsian mutation" remains unknown, however.

The Characterization of Tetracycline Resistant Mutants

As mentioned above, the most valuable mutant for chemotherapeutic purposes is the one that could be producing an inhibitor of the HIV protease. The difficulty is to determine which of the mutants might be in this category and to identify it among other mutants such as those which have reverted the *tet* gene or have inactivated the protease. Thus, we have outlined a scheme for the genetic determination of such mutants (See Figure 2). In our initial experiments, plasmids were isolated from 20 Tet^r mutants and re-transformed into naive HB101 (Step B). The resulting transformants were selected for ampicillin and chloramphenicol resistance and screened for Tet resistance. From the original 20 Tet^r mutants, only one had plasmids that conferred tetracycline resistance. These results indicate that the plasmids isolated from the other 19 Tet^r mutants were not altered and suggest that the original mutations which conferred resistance to tetracycline were not the result of a chromosomal mutation. This is probably due to the fact that both the HIV protease and the Tet protein are on multicopy plasmids and it would be rare to identify a mutation in these plasmid genes. These results suggested that plasmid derived mutants would not present a major difficulty in the screen but that chromosomal mutations leading to Tet resistance, which we did not expect to find, could be the problem.

To exploit the power of *E. coli* genetics, we have designed an *in vivo* co-cultivation assay for the initial characterization of Tet resistant mutants to identify those that might be producing potential inhibitors of the HIV protease (Figure 2, Step C). First, we constructed a new testor strain that would contain the HIV protease and the target Tet protein in a Lac$^-$ background (JM109). On IPTG/X-gal plates this strain (RG539) is white whereas the testor strain (RG533) in HB101 (Lac$^+$) is blue. Tet resistant mutants (putative synthesizers of an HIV protease inhibitor) were then selected on tetracycline plates using strain RG539. The resulting mutants (Tet^r, white) were screened for the synthesis of an inhibitor of the HIV protease using a co-cultivation assay with the original testor strain RG533 (Tet^s, blue). This co-cultivation assay is shown in Figure 3. Only Tet^r, white colonies, that secrete an inhibitor of the HIV protease (or somehow interfere with tetracycline uptake) will allow the blue, tetracycline sensitive test strain to grow. Thus far, we have identified 20 such Tet^r mutants that can restore the ability of the blue, Tet^s testor colonies to grow.

One such Tet^r mutant that could be elaborating an inhibitor of the HIV protease has been characterized further and is shown in Figure 4. This mutant (#11) had been isolated after five days of selection in tetracycline. In row #1, the growth of the blue Tet sensitive testor colonies is readily seen on the white background of the Tet resistant mutant. These blue colonies have been isolated and are still Tet sensitive,

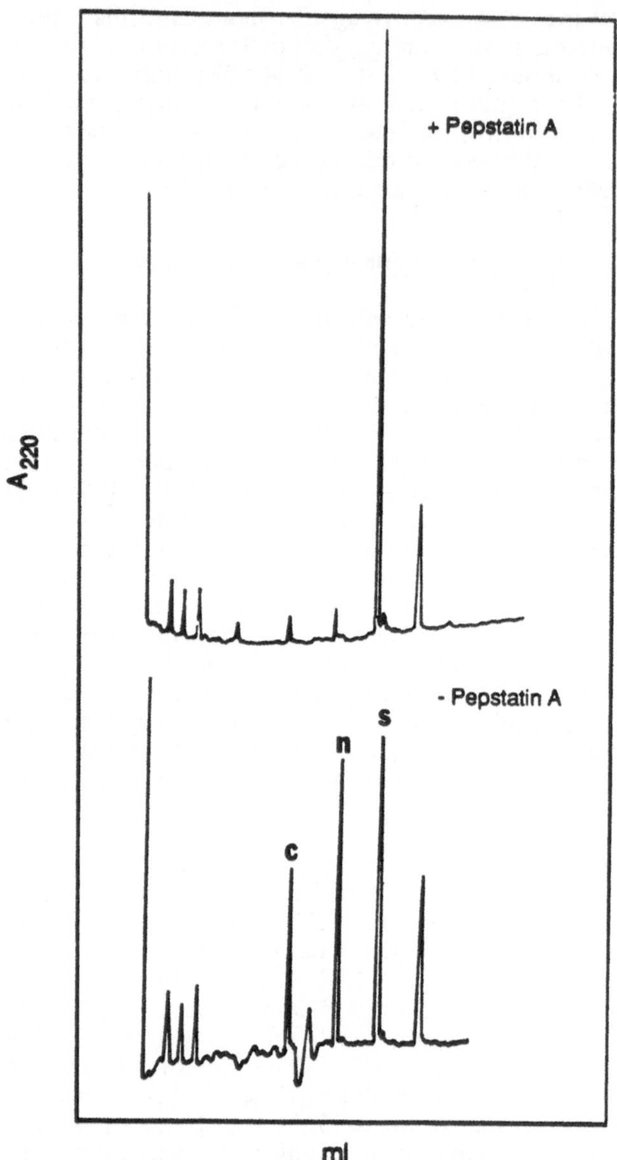

Figure 5. Biochemical assay of the HIV protease. The HIV protease assay kit was purchased from Bachem Bioscience and used according to the manufacturer's instructions. Reaction was for 1 hr at 37° and the products were analyzed using a linear gradient of 10-100% (A=0.1% TFAin H_2O; B=0.1% TFA in 50% ACN) in 20 minutes at 1 ml/min on an LKB C-18 column (TSK ODS-120T; 4.6 x 250 mm) and monitoring the absorbance at 220 nm. The substrate(s) eluted at 16 min; the carboxy terminal fragment (c) eluted at 12 min; the amino terminal fragment (n) eluted at 14 min.

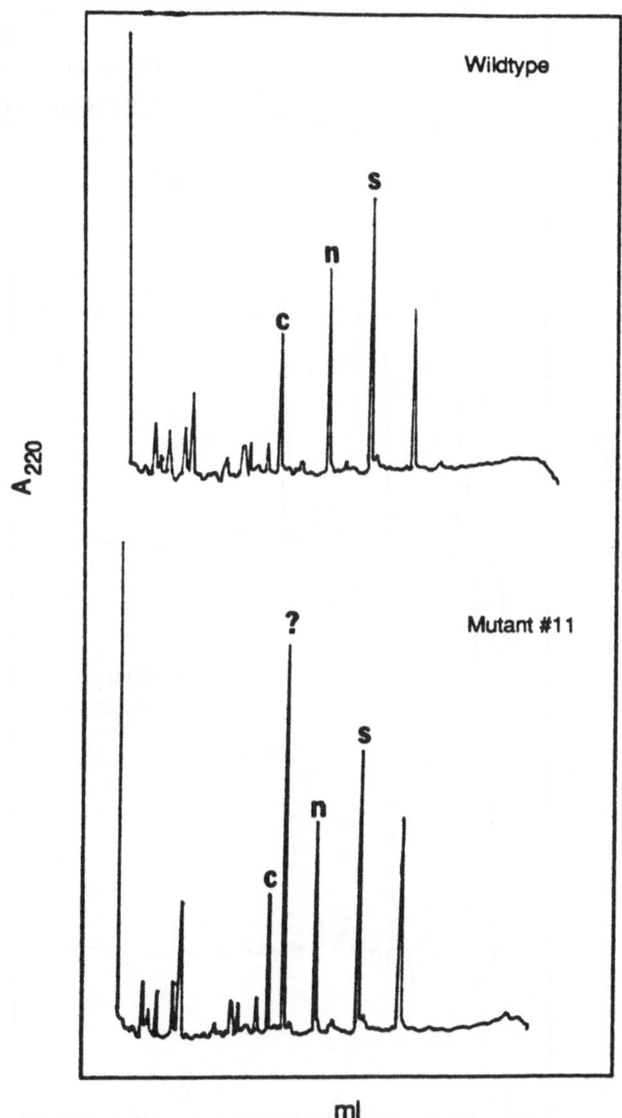

Figure 6. Assay for the presence of a protease inhibitor from mutant #11. Reaction conditions and HPLC analysis were the same as that described in Figure 5 except that each assay contained media from an overnight culture of mutant #11 or from its parent strain (Wildtype).

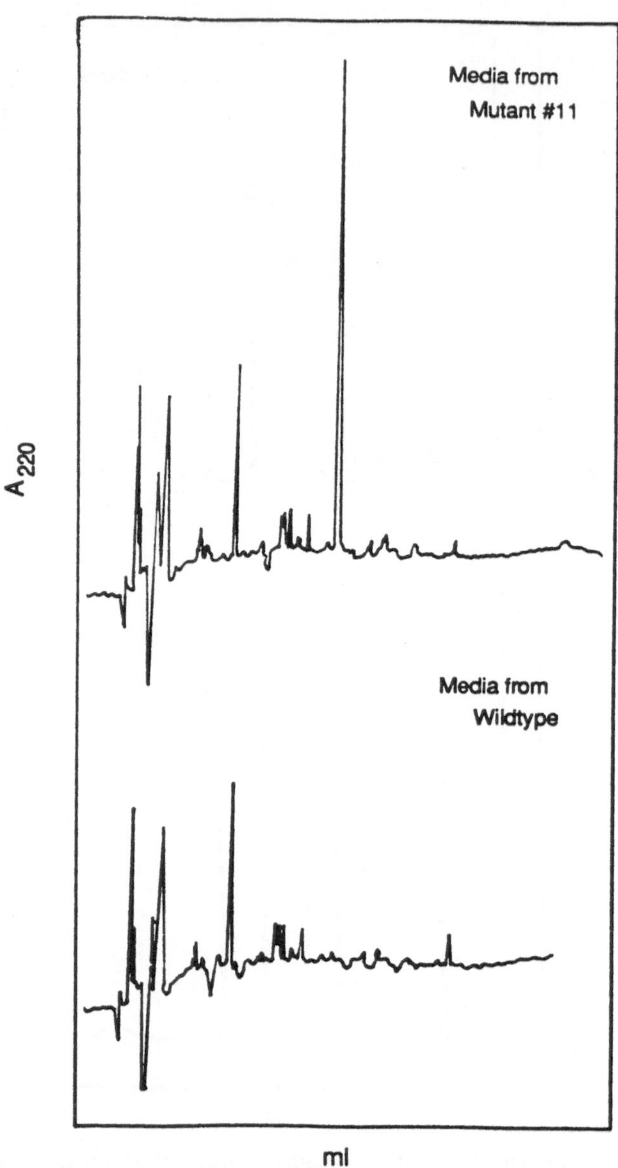

Figure 7. HPLC analysis of the media from an overnight culture of mutant #11 of wildtype. Media from either mutant #11 or wildtype were analyzed by HPLC as described in Figure 5 but without prior digestion with the HIV protease.

there is no detectable inhibition of the HIV protease (58% and 59% digestion, respectively). However, in the case of mutant #11 an additional peak is present that is not observed in the wildtype assay. To test whether or not this new peak was the result of degradation by the protease, the media from either mutant #11 or wildtype were chromatographed without prior enzyme treatment. As shown in Figure 7, this new peak is not the result of HIV protease degradation but is only found in the media from mutant #11 and not in the parent strain. Upon further analysis, this peak has now been identified as tryptophan. That is, mutant #11 is Tet^r because it inhibits the expression of the HIV protease through the production of tryptophan since the protease is under the regulation of the Trp promoter. Presumably, this is a cis mutation in the operator of the Trp operon that prevents the Trp repressor from binding and allows the overproduction of tryptophan which can then activate the Trp repressor and turn off the transcription of the protease. Of the 20 mutants which we have identified as allowing the HIV protease strain to grow in the co-cultivation assay, 19 have tryptophan in the media and are presumably this class of mutation. The other one, which does not have tryptophan in the media, is currently under investigation.

Concluding Remarks

Bacteria have been the source of numerous important therapeutic compounds. Perhaps it is not too fanciful to think that by application of the powerful tool of bacterial genetics mutant bacteria can be selected that will inhibit foreign proteins which they are expressing. Ironically, most of the efforts to produce novel antiviral compounds have relied on the rational design of inhibitors based on the biochemical and structural properties of the protein (Dreyer et al., 1989; Kholberg et al., 1991; McQuade et al., 1990; Miller et al., 1989; Moore et al., 1989; Pauwels et al., 1990; Roberts et al., 1990). Here, by selecting for mutants, we have simply asked the bacteria to "irrationally" design an inhibitor of the HIV protease. The results presented here demonstrate that the bacteria have solved the problem of growth in tetracycline by producing an inhibitor of the expression of the HIV protease, i.e. tryptophan. If a bacterially produced inhibitor of the biochemical activity of the protease can be found, then perhaps many of the problems with the rationally designed compounds (solubility, stability, etc.) would also have solved by the bacteria. Even if the putative inhibitor has no therapeutic value, we will be justified in continuing to search for one that does. Changing the bacterial strain to a Trp auxotroph or to a different bacterial genera, or even to yeast could afford the identification of alternative ways that microorganisms can devise to inhibit the HIV protease. The potential of this method is only limited by our own imagination to develop future assays to uncover even more important uses for microorganisms and the compounds they produce.

ACKNOWLEDGEMENTS

This work was supported by the Dean's Overage Research Program from Thomas Jefferson University and the W.W. Smith Charitable Trust.

REFERENCES

Ames, B.N., McCann, J., and Yamasaki, E., 1975, Methods for detecting carcinogens and mutagens with the *Salmonella*/mammalian-microsome mutagenicity test, Mutation Research, 31:347-364.

Ashby, J., and Tennant R.W., 1988, Chemical structure, Salmonella mutagenicity and extent of carcinogenicity as indicators of genotoxic carcinogens among 222 chemicals tested on rodents by the U.S. NCI/NTP, Mutation Research 204:17-115.

Backman, K. and Boyer, H.W., 1983, Tetracycline resistance determined by pBR322 is mediated by 1 polypeptide, Gene 26:197-203.

Barany, F., 1985, Two-codon insertion mutagenesis of plasmid genes by using single-stranded hexameric oligonucleotides, Proc. Natl. Acad. Sci. USA 82:4202-4206.

Baum, E.Z., Bebernitz, G.Z., and Gluzman, Y., 1990a, Beta galactosidase containing a human immunodeficiency virus protease cleavage site is cleaved and inactivated by human immunodeficiency protease, Proc. Natl. Acad. Sci. USA 87:10023-10027.

Baum, E.Z., Bebernitz, G.A., and Gluzman, Y., 1990b, Isolation of mutants of human immunodeficiency virus protease based on the toxicity of the enzyme in *Escherichia coli*, Proc. Natl. Acad.Sci. USA 87:5573-5577.

Bjornsti, M.A., Benedetti, G.A., Viglianti, A., and Wang, J.C., 1989, Expression of human DNA topoisomerase I in yeast cells lacking yeast DNA topoisomerase I: Restoration of sensitivity of the cells to the antitumor drug camptothecin, Cancer Research 49:6318-6323.

Block, T.M., and Grafstrom R.H., 1990, Novel bacteriological assay for the detection of potential antiviral agents, Antimicrob. Agents and Chemo. 34:2337-2341.

Borel, J.F., Feurer, C., Magnee, C., and Stahelin, H., 1977, Effects of the new anti-lymphocytic peptide cyclosporin A in animals, Immunol. 32:1017-1025.

Cairns, J., Overbaugh, J., and Miller, S., 1988, The origin of mutants, Nature 335:142-145.

Curiale, M.S., McMurry L.M., and Levy, S.B., 1984, Intracistronic complementation of the tetracycline resistance membrane protein of Tn10, J. Bacteriol. 157:211-217.

Cwirla, S.E., Peters, E.A., Barrett, R.W., and Dower, W.J., 1990, Peptides on phage: A vast library of peptides for identifying ligands, Proc. Natl. Acad. Sci. USA 87:6378-6382.

Debouck, C., Gorniak, J.G., Strickler, J.E., Meek, T.D., Metcalf, B.W., and Rosenberg, M., 1987, Human immunodeficiency virus protease expressed in *Escherichia coli* exhibits autoprocessing and specific maturation of the *gag* precursor, Proc. Natl. Acad.Sci. USA 84:8903-8906.

DeClercq, KE., 1990, Targets and strategies for the antiviral chemotherapy of AIDS, TIPS 11:198-205.

Dreyer G.B., Metcalf, B.W., Tomaszek, T.A., Jr., Carr, T.J., Chandler, A.C., III, Hyland, L., Fakhouryk, S.A., Magnard, V.W., Moore, M.L., Strickler, J.F., Debouck, C., and Meek, T.D.1989, Inhibition of HIV-1 protease in vitro: rational design of substrate analogue inhibitors, Proc. Natl.Acad. Sci. USA 86:9752-9756.

Evnin, L.B., Vazquez, J.R., and Craik, C.S., 1990, Substrate specificity of trypsin investigated by using a genetic selection, Proc. Natl. Acad. Sci. USA 87:6659-6663.

Hall, B., 1990, Spontaneous point mutations that occur more often when advantageous than when neutral, Genetics 126:5-16.

Hansen, J., Billich, S., Schulze, T., Sukrow, S., and Moelling, K., 1988, Partial purification and substrate analysis of bacterially expressed HIV protease by means of monoclonal antibody, EMBO J. 7:1785-1791.

Harley, C.B., Laurie, J., Betlack, M., Crea, R., Boyer H.W., and Hedgpeth, J., 1988, Transcription initiation a the tet promoter and effect of mutations, Nucleic Acids Res. 16:7269-7285.

Harley, C.B., Laurie, J., Boyer, H.W., and Hedgpeth, J., 1990, Reiterative copying by the *E. coli* RNA polymerase during transcription initiation of mutant pBR322 tet promoters, Nucleic Acids Res. 18:547-552.

Huse, W.D., Sastry, L., Iverson, S.A., Kong, A.S., Alting, S., Mees, M., Burton, D.R., Benkovic, S.J., and Lerner, R.A.,1989, Generation of a large combinational library of the immunoglobulin repertoire in phage lambda, Science 246:1275-1281.

Ivanoff, L.A., Towatari, T., Ray, J., Korant, B.D., and Petteway, S.R., 1986, Expression and site specific mutagenesis of the poliovirus 3C protease in *Escherichia coli*, Proc. Natl. Acad. Sci. USA 83:5392-5396.

Kadonaga, J.T., Carner, K.R., Masiarz, F.R., and Tijan, R., 1987, Isolation of cDNA encoding transcription factor Sp1 and functional analysis of the DNA binding domain, Cell 51:1079-1090.

Kholberg, R., 1991, Critics call for a smarter way to screen for drugs, J. NIH Res. 3:25-26.

Kohl, N.E., Emini, E.A., Schleif, W.A., Davis, L.J., Heimbach, J.C., Dixon, R.A.F., Scolnick, E.M., and Sigal, I.S., 1988, Active immunodeficiency virus protease is required for viral infectivity, Proc. Natl. Acad. Sci. USA 85:4686-4690.

Korant, B., 1990, *AIDS Research and Reference Reagent Program Catalog*, U.S. Department of Health and Human Services, p.51 (January).

Krausslich, H.G., Schneider, H., Zybarth, G., Carter, C.A., and Wimmer, E., 1988, Processing of in vitro synthesized *gag* precursor proteins of human immunodeficiency virus (HIV) type-1 by HIV protease generated in *Escherichia coli*, J. Virol. 62:4393-397.

Legrice, S.F.J., Mills, J., and Mous, J., 1988, Active site mutagenesis of the AIDS virus protease and its alleviation by trans complementation, EMBO J. 7:2547-2553.

Marshall, B., Morrissey, S., Flynn, P., and Levy, S.B., 1986, A new tetracycline resistant determinant, class E isolated from Enterobacteriaceae, Gene 50:111-117.

McCann, J., Choi, E., Yamasaki, E., and Ames, B.N., 1975,Detection of carcinogens as mutagens in the Salmonella/microsome test: Assay of 300 chemicals, Proc. Natl. Acad. Sci. USA 72:5135-5139.

McMurry, L., Petrucci, R.E., and Levy, S.B., 1980, Active efflux of tetracycline encoded by four genetically different tetracycline resistance determinants of *Escherichia coli*, Proc. Natl. Acad. Sci. USA 77:3974-3977.

McQuade, T.J., Tomasselli, A.G., Liu, L., Karacostas V., Moss, B., Sawyer, T.K., Heinrikson, R.L., Tarpley, W.G., 1990, A synthetic HIV-protease inhibitor with antiviral activity arrests HIV-like particle maturation, Science 247:454-456.

Middleman, E., Luce, J., and Frei, E., 1971, Clinical trials with adriamycin, Cancer 28:844-850.

Miller, M., Schneider, J., Sathyanarayana, B.K., Toth, M.V., Markshall, G.R., Clawson, L., Selk, L., Kent, S.B.H., and Wlodawer, A.,1989, Structure of complex of synthetic HIV-1 protease with a substrate-based inhibitor at 2.3 A resolution, Science 246:1149-1151.

Mitsuya, H., Yarchoan, R., and Broder, S., 1990, Molecular targets for AIDS therapy, Science 249:1533-1544.

Moore, M.L., Bryan, W.M., Fakhoury, S.A., and Maagard, V.M., Huffman, W.F., Dayton, B.D., Meek, T.D., Hyland, L., Dreyer, G.B., Metcalf, B.W., Strickler, J.E., Gorniak, J.G., and Debouck, C., 1989, Peptide substrates and inhibitors of the HIV-1 protease, Biochem. Biophys. Res. Comm. 159:420-425.

Pauwels, R., Andries, K., Desmyter, J., Schols, D., Kukla, M.J., Breslin, H.J., Raeymaeckers, A., Van Gelder, J., Woestenborghs, R., Heykants, J., Schellekens, K., Janssen, M.A.C., DeClerq, E., and Janssen, P.A.J., 1990, Potent and selective inhibition of HIV-1 replication in vitro by a novel series of TIBO derivatives, Nature 343:470-474.

Roberts, N.A., Martin, J.A., Kinchington, D., Broadhurst, A.V., Craig, J.C., Duncan, I.B., Galpin, S.A., Handa, B.K., Krohn, J.K.A., Lambert, R.W., Merrett, J.H., Mills, J.S., Parkes, K.B.B., Redshaw, S., Ritchie, A.J., Taylor, D.L., Thomas, G.J., Machin, P.J., 1990, Rational design of peptide-based HIV proteinase inhibitors, Science 248:358-361.

Rubin, R.A., and Levy, S., 1990, Interdomain hybrid Tet proteins confer tetracycline resistance only when they are derived from closely related members of the *tet* gene family, J. Bacteriol. 172:2303-2312.

Seelmeir, S., Schmidt, H., Turk, V., and von der Helm, K., 1988, Human immunodeficiency virus has an aspartic-type protease that can be inhibited by pepstatin A, Proc. Natl. Acad. Sci. USA 85:6612-6616.
Sidwell, R.W., Dixon, G.J., Schabel, F.M., 1968, Antiviral activity of 9-beta-D-arabinofuranosyl adenine, Antimicrob. Agent. Chemo. 8:148-154.
Yang, H.L., Zubay, kG., and Levy, S., 1976, Synthesis of a R plasmid associated with tetracycline resistance is negatively regulated, Proc. Natl. Acad. Sci. USA 73:1509-1512.

ANTIVIRAL AGENTS FROM NOVEL MARINE AND TERRESTRIAL SOURCES

Kenneth L. Rinehart

Department of Chemistry, University of Illinois, Urbana, Illinois 61801

Pharmaceutical products are derived from only two sources--natural products and organic synthesis; indeed, natural products have proven a cornucopia of drugs through the years. Systematic searches for antiviral agents in nature, however, are both rare and recent. Our own extensive efforts to identify natural products with antiviral activity began with a one-month expedition to the Western Caribbean in 1978 on board the National Science Foundation's *R/V Alpha Helix* (*A*lpha *H*elix *C*aribbean *E*xpedition, AHCE 1978).[1] The expedition started in Panama, extended to Cozumel, Mexico, and visited roughly 20 collecting sites en route. The focus was on Belize, the site of the largest reef in the Western Hemisphere. During the expedition we collected somewhat over 500 different species and a total of about 1000 samples. The scientific crew of this expedition included Dr. Robert G. Hughes, Jr., Roswell Park Memorial Institute, Buffalo, NY, who carried out antiviral assays on shipboard by using a standard *Herpes simplex* virus, type 1 (HSV-1), assay, with the virus being grown in monkey kidney, CV-1, cells.[2] This allowed us to identify not only antiviral, but also cytotoxic species. Because we had no prior information on active species, our goal was to collect by SCUBA every marine invertebrate and alga we could find and to assay every sample for antiviral activity. Then, while on location, we planned to collect more of the bioactive species.

On-site bioassays have several advantages:[3] active species can be re-collected, the samples are fresh, and the freezer is filled with active materials. The antiviral assay in AHCE 1978 was HSV-1, but in other expeditions we have used *Vesicular stomatitis* virus (VSV) as an example of an RNA virus and, more recently, A59, a murine corona virus. The AHCE 1978 results in Table 1 show rather high percentages of antiviral activity in a few phyla, including the Chordata (tunicates) and the Porifera (sponges).

Antiviral Species from AHCE 1978

Table 2 contains a priority list of antiviral species; the most active sample (++) was a tunicate, but several other extracts were active (+). The Upjohn Company's Dr. Harold E. Renis provided secondary testing of these extracts against a variety of viruses, and we were disappointed to find that many of the samples which were positive on-site, including the most active, proved negative in the secondary testing. However, we later found that fresh samples from re-collections were again active, and we were subsequently able to isolate the active materials, as detailed below. These experiences confirmed the wisdom of on-site testing. For the isolation of the antiviral compounds, we have sometimes employed a thin-layer chromatography/bioautography system to follow the purification.[4]

Sceptrins, Oxysceptrins, Ageliferins

The *Agelas* sponge genus gave us three different samples in Table 2 (#'s 100, 126, 319), from which we were able to isolate a family of antiviral compounds.[5-8] Sceptrin (**2**), the first member of the series, had been isolated (with no indication of antiviral activity) by the Faulkner

Table 1. Antiviral Activity and Cytotoxicity in Phyla Assayed during the Alpha Helix Caribbean Expedition 1978

	% Species Active (Number of Species Examined)	
Phylum	HSV-1[a]	CV-1[b]
Porifera	14 (180)	62 (186)
Cnidaria	17 (69)	56 (70)
Ectoprocta	0 (1)	0 (1)
Mollusca	0 (21)	33 (21)
Annelida	0 (3)	0 (3)
Arthropoda	0 (6)	0 (6)
Echinodermata	16 (36)	72 (36)
Chordata	23 (26)	70 (27)
Cyanophyta	100 (5)	80 (5)
Chlorophyta	7 (42)	36 (42)
Phaeophyta	25 (19)	50 (19)
Rhodophyta	17 (42)	43 (42)
Tracheophyta	33 (3)	0 (3)

[a]Inhibiting *Herpes simplex* virus, type 1, at ≤200 μg/disk. [b]Cytotoxic to monkey kidney cells at ≤200 μg/disk.

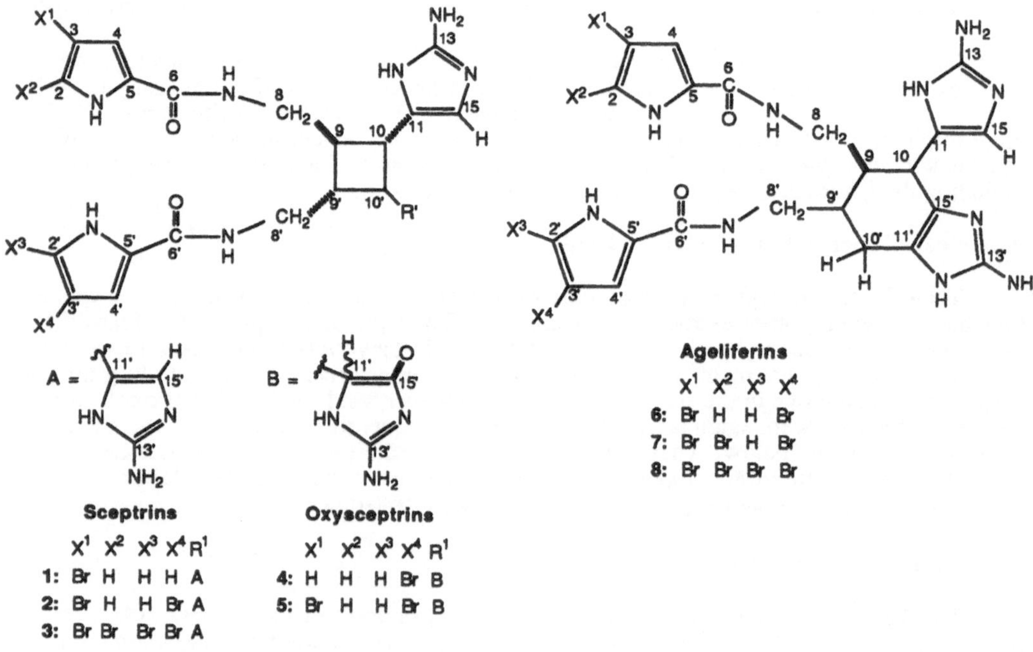

Sceptrins
X1 X2 X3 X4 R1
1: Br H H H A
2: Br H H Br A
3: Br Br Br Br A
Oxysceptrins
X1 X2 X3 X4 R1
4: H H H Br B
5: Br H H Br B
Ageliferins
X1 X2 X3 X4
6: Br H H Br
7: Br Br H Br
8: Br Br Br Br

Table 2. Priority Lists of Species from AHCE 1978 with Antiviral Activity or Cytotoxicity

Antiviral Activity, Shipboard[a]	AHCE Sample	Phylum	Secondary Testing[b]	Cytotoxicity Index, Shipboard[c]	AHCE Sample	Phylum	Secondary Cytotoxicity (ID_{50}, µg/mL)[d]
→ ++	#553	Chordata	–	318	#418	Mollusca	0.30
→ +	100	Porifera	–	283	650	Porifera	0.17
+	175	Chordata	–	136	639	Porifera	0.094
→ +	55	Chordata	+	116	265	Chordata	NT[e]
+	371	Porifera	–	103	721	Porifera	NT
→ +	319	Porifera	–	98	485	Porifera	3.7
+	112	Cnidaria	–	95	59	Porifera	0.16
→ +	126	Porifera	–	89	754	Porifera	NT
+	1058	Rhodophyta	–	89	150	Cnidaria	0.35
+	1008	Tracheophyta	–	85	303	Cnidaria	NT
→ +	755	Chordata	+	80	751	Porifera	NT
+	441	Echinodermata	–	→ 80	755	Chordata	0.90
+	177	Echinodermata	–	79	421	Echinodermata	NT
+	84	Echinodermata	–	76	642	Porifera	NT
+	155b	Cnidaria	NT	72	761	Porifera	NT
+	169	Cnidaria	–	72	30	Cnidaria	3.0
+	463	Porifera	–	72	1020	Rhodophyta	19.0
→ +	101	Porifera	+	67	40	Porifera	3.2
+	754	Porifera	+	67	82	Cnidaria	3.4
+	1019	Cyanophyta	+	63	32	Cnidaria	3.8
+	1173	Cyanophyta	+	63	27	Cnidaria	0.70
+	1072	Rhodophyta	–	63	1028	Rhodophyta	7.6
+	160	Chordata	–	63	641	Porifera	>1.0
				62	156	Cnidaria	0.83
				62	179	Porifera	>2.5
				62	181	Porifera	0.28

[a]Activity against Herpes simplex virus, type 1; from 100 µL of a 20-mL methanol-toluene (3:1) extract of 2 g of sample; ++ = only a few viral plaques formed, + = definite inhibition of viral plaque formation. [b]Upjohn screen; + = active against one or more viruses at 1 mg/mL. [c]Zone of inhibition of CV-1 (monkey kidney) cells, extrapolated to 100 µL of a 20-mL methanol-toluene (3:1) extract of 2 g of sample. [d]Upjohn screen; L1210 cells. [e]NT = not tested.

#101--*Ptilocaulis* aff. *P. spiculifer*; #100, 126, 319--*Agelas* species; #553--*Eudistoma olivaceum*; #55, 755--*Trididemnum solidum.*

group at the Scripps Institution of Oceanography,[9] but the rest of the compounds were new, with structures assigned by Drs. Paul A. Keifer and Moustapha E. S. Koker in our laboratory. Some were brominated and/or oxygenated analogues of sceptrin, while the ageliferins had a six-membered central ring. All of these compounds had antiviral properties, but they were not strongly antiviral. Sceptrin and ageliferin (**6**) and their debromo-, bromo-, and dibromo-analogues were active against HSV-1 at 20 µg/disk but against VSV only at 100 µg/disk, and oxysceptrin (**5**) and debromooxysceptrin (**4**) were less active. In general, we regard compounds as being strongly antiviral when plaque formation is inhibited at less than 1 µg/disk, while those of intermediate activity inhibit at 1-10 µg/disk, and those of weak activity inhibit at greater than 10 µg/disk. While the *Agelas* compounds are only weakly active, they are so abundant in the sponges that the extracts test positive.

Eudistomins

The tunicate *Eudistoma olivaceum* provided our most active antiviral extract from AHCE 1978 (Sample #553 in Table 2), as noted above. This tunicate grows on mangrove roots at depths up to 1 meter and has been a very frustrating species because it is not abundant; it was collected only once during AHCE 1978. Subsequently, we have found samples in Florida, Puerto Rico, and other Caribbean locations, but never in great abundance. The compounds we isolated from *E. olivaceum*, the eudistomins, all have a β-carboline nucleus, as shown here.[10,11] The simplest eudistomins have no substituents attached to the pyridine ring; others

→D: R = Br, R^{I} = OH, R^{II} = H
J: R = H, R^{I} = OH, R^{II} = Br
→N: R = R^{II} = H, R^{I} = Br
→O: R = R^{I} = H, R^{II} = Br

C: R = R^{III} = H, R^{I} = OH, R^{II} = Br
E: R = Br, R^{I} = OH, R^{II} = R^{III} = H
F: R = H, R^{I} = OH, R^{II} = Br, R^{III} = $C_2H_3O_2$
K: R = R^{I} = R^{III} = H, R^{II} = Br
L: R = R^{II} = R^{III} = H, R^{I} = Br

A: R = OH, R^{I} = Br
→ M: R = OH, R^{I} = H

G: R = H, R^{I} = Br
→H: R = Br, R^{I} = H
→I: R = R^{I} = H
P: R = OH, R^{I} = Br
→Q: R = OH, R^{I} = H

The Eudistomins

Table 3. *In vitro* Antiviral Activity of Eudistomins and Derivatives

Eudistomin	HSV-1[a]	ng/12.5-mm disk
D	+	500
D (synth)	+	500
D-Ac_2	+	500
D-Ac (synth)	±	1250
7-Br-D (synth)	+	500
J	±	1000
J-Ac_2	±	500
N (synth)	–	1250
N + O	+	500
A	–	500
A-Ac	+	1000
M-Ac	±	500
G	±	500
H	+	500
I	±	500
P	+	500
Q	±	500
B	–	500
C	+++	50
	++	25
	+	10
	–	5
C-Ac_2	+	1000
E	+++	50
	+++	25
	±	5
K	+	250
L	+	100

[a]+++, ++, +, ±, – = complete → no inhibition.

are pyrrolyl- or pyrrolinyl-substituted on the pyridine ring. The initial structural work was carried out by Drs. Gary C. Harbour and Jun'ichi Kobayashi in our laboratory. The arrows refer to compounds which we have subsequently synthesized, largely through the efforts of Dr. Jeremy Gilmore, and those compounds are available in reasonable quantities.

The most interesting compounds in the series, however, are highly modified β-carbolines with a reduced tetrahydropyridine ring and a seven-membered oxathiazepine ring containing nitrogen, oxygen, and sulfur. The oxathiazepine ring system was unknown in nature, had never been prepared, and is rather difficult to synthesize. At least four groups worked over a 4-year period on the system, and they have published the syntheses of compounds in this series only within the last two years.[12-15]

It is unfortunate that larger quantities of the oxathiazepine-containing eudistomins have not been available because these are by far the most active of the eudistomins, as shown by the in vitro activities in Table 3. All compounds were tested against HSV-1 in CV-1 cells. Some of the simpler, aromatic β-carbolines are reasonable antiviral agents, with inhibition at 500 ng/disk. However, the eudistomins with oxathiazepine rings can be remarkably active; eudistomins C and E, both with hydroxyl and bromine substituents on the benzene ring, inhibit at about the 5- to 10-ng level.

In vivo studies on the eudistomins have also been carried out by Dr. Renis at Upjohn. In the mouse vaginal assay, most of the mice die if not treated, but about 80% can be saved over the 21 days of the assay by topical treatment with eudistomin C (0.1 mg/mL) and a slightly smaller percentage with eudistomin E. Much lower in vivo activity was shown by simpler β-carbolines, without the oxathiazepine ring. Additional studies seem warranted on the entire class of compounds now that they are synthetically available.

Didemnins

A second tunicate which produces antiviral compounds, *Trididemnum solidum*, looks like a pancake growing on corals, sponges, and other fixed objects from 30 to 120 feet in depth. The compounds produced by *T. solidum* are called didemnins, shown here.

The structures of didemnins A-E were assigned by Dr. James B. Gloer,[16,17] those of didemnins X and Y by Ryuichi Sakai,[18,19] and that of dehydrodidemnin B by Dr. Anna Lithgow-Bertelloni,[20] all in our laboratory. They are interesting compounds structurally, containing a number of standard amino acids, together with the unusual dimethyltyrosine and *N*-methyl-D-leucine and the most unusual isostatine.[21,22] Isostatine is closely related to statine, a component of pepstatin, but it is derived biosynthetically from isoleucine and acetate units instead of from leucine and acetate units as found in statine. The other highly unusual component is γ-hydroxyisovalerylpropionic acid (Hip). The most active didemnin in earlier tests, didemnin B,[23,24] is about 10 times as active as didemnin A and differs from it in having an extra proline and a lactic acid unit in the side chain.

Didemnin E is about as active as didemnin B, and dehydrodidemnin B is more active. Nevertheless, didemnin B has been studied by far the most, and most of the data that follow come from didemnin B testing.[25,26] Didemnin B is active as a cytotoxic and antineoplastic agent as well as being antiviral, and its in vivo antitumor activity led it through toxicity studies into Phase I and Phase II clinical trials as a potential anticancer agent.[27] It has been in Phase II trials for about three years, and results may be released in another year. Didemnin B is also very active as an immunosuppressive agent. It is approximately 100 to 1000 times as active as cyclosporin A in most assays which are comparable, such as inhibition of T-cell mitogenesis and the mixed lymphocyte reaction.[28,29] Immunomodulation studies are continuing in a number of laboratories.

Returning to the antiviral activity, we can note that different didemnins have somewhat different activity patterns. Table 4 shows Dr. Renis's results, where the number before the diagonal is a cytotoxicity measure, that after it an antiviral measure. We have been trying to improve the ratio, and it is clear that some analogues and derivatives of didemnins have

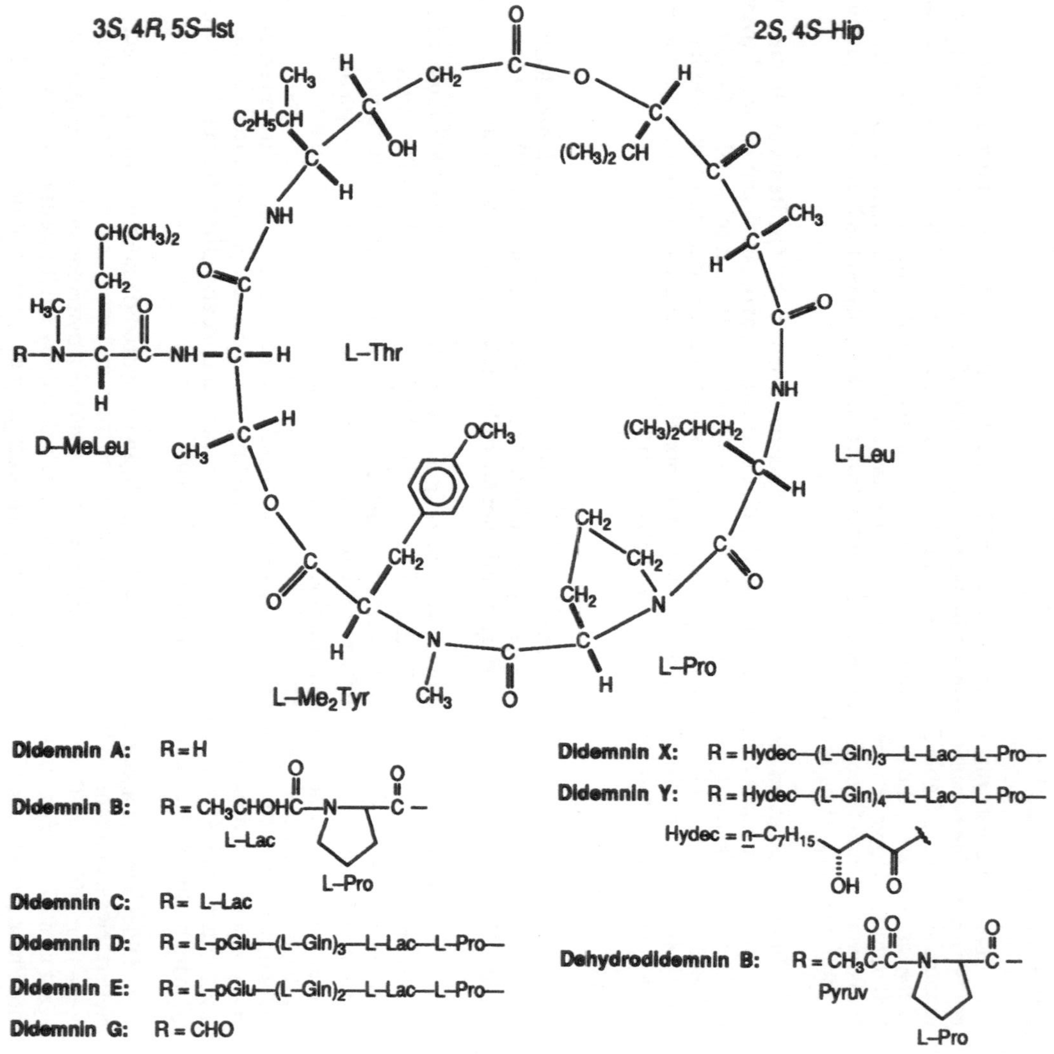

3*S*, 4*R*, 5*S*–Ist

2*S*, 4*S*–Hip

D–MeLeu

L–Thr

L–Leu

L–Me$_2$Tyr

L–Pro

Didemnin A: R = H

Didemnin B: R = $CH_3CHOHC(=O)$–N(L–Pro)–C(=O)– (L–Lac, L–Pro)

Didemnin C: R = L–Lac

Didemnin D: R = L–pGlu–(L–Gln)$_3$–L–Lac–L–Pro–

Didemnin E: R = L–pGlu–(L–Gln)$_2$–L–Lac–L–Pro–

Didemnin G: R = CHO

Didemnin X: R = Hydec–(L–Gln)$_3$–L–Lac–L–Pro–

Didemnin Y: R = Hydec–(L–Gln)$_4$–L–Lac–L–Pro–

Hydec = n–C_7H_{15}–CH(OH)–CH_2–C(=O)–

Dehydrodidemnin B: R = $CH_3C(=O)C(=O)$–N(L–Pro)–C(=O)– (Pyruv, L–Pro)

Table 4. *In Vitro* Bioactivities of the Didemnins and Their Derivatives

	RNA Viruses[a]				DNA Viruses[a]			L1210 Cells[b]	
Compound	PR8	COE	HA-1	E.R.	HSV-1	HSV-2	Vacc	ID_{50}	ID_{90}
Didemnin A						2/3		0.019	0.056
Didemnin B	4/0	4/0	4/0	4/0	4/0	3/3	4/0	0.0011	0.0049
Synthetic								0.0018	0.0135
Didemnin C	4/0	3/3	4/4	4/0	1/4	2/3	1/4	0.011	0.019
Didemnin D	4/0	4/0	4/0	4/4	4/0	4/0	4/0	0.0065	0.016
Didemnin E	4/0	4/0	4/0	4/4	4/0	4/0	4/0	0.0051	0.013
Didemnin G								0.006	0.038
Didemnin X								0.0048	0.017
Didemnin Y								0.0048	0.021
Methylene didemnin A	0/0	0/2	0/4	0/4	0/4	0/4	0/4	0.0065	0.023
N-Acetyldidemnin A								0.0024	0.007
Diacetyldidemnin A	0/0	0/3	0/4	0/4	1/4	1/4	1/4	0.015	0.052
Dihydrodidemnin A								0.52	>1
Nordidemnin B	4/0	4/0	4/0	4/0	4/4	4/4	3/4	0.0078	0.019
Diacetyldidemnin B	3/0	3/3	1/0	2/4	3/4	3/4	3/4	0.0016	0.0036
Prolyl-didemnin A								0.014	0.076

[a]Cytotoxicity/antiviral activity, 1 = 1-10, 2 = 10-20, 3 = 20-30, 4 = 30-40 mm zone of inhibition; PR8, influenza virus; COE, Coxsackie A21 virus; HA-1, parainfluenza-3 virus; E.R., equine rhinovirus; HSV-1, HSV-2, *Herpes simplex* virus, types 1 and 2; Vacc, vaccinia virus. [b]μg/mL.

relatively greater antiviral activity. Some of those compounds should now be tested in vivo for antiviral activity.

Didemnins are very potent inhibitors of viruses in vitro. Among DNA viruses, viral titer reductions of approximately four orders of magnitude for HSV-1 and HSV-2 can be effected at 5 μg/mL (Figure 1). Didemnins are also effective in vitro against Dengue virus (Figure 2), as shown by Dr. Edmundo N. Kraiselburd at the University of Puerto Rico.[30] Dr. Erik DeClercq at the University of Leuven, Belgium, has demonstrated that they are very potent against *Varicella zoster* and cytomegalovirus (Table 5); however, the selectivity index is not high. This illustrates the difficulties in using the current didemnins as antiviral agents per se.

Table 5. DNA Antiviral Activity of Didemnin B[a]

	ID_{50}, μg/mL (PFU reduction)[b,c]	Selectivity index	ID_{50}, μg/mL (CPB reduction)[b,d]	Selectivity index
Varicella zoster				
YS (TK+)	0.02	15		
Oha (TK+)	0.02	15		
7-1 (TK−)	0.04	7.5		
YSR (TK−)	0.02	15		
Cytomegalovirus				
Davis	0.02	15	0.02	15
AD-169	0.06	5	0.07	4

[a]E. DeClercq, T. Sakuma, University of Leuven. [b]HEL (human embryonic lung) cells. [c]20 plaque forming units (PFU)/well. [d]Cytopathogenic effect.

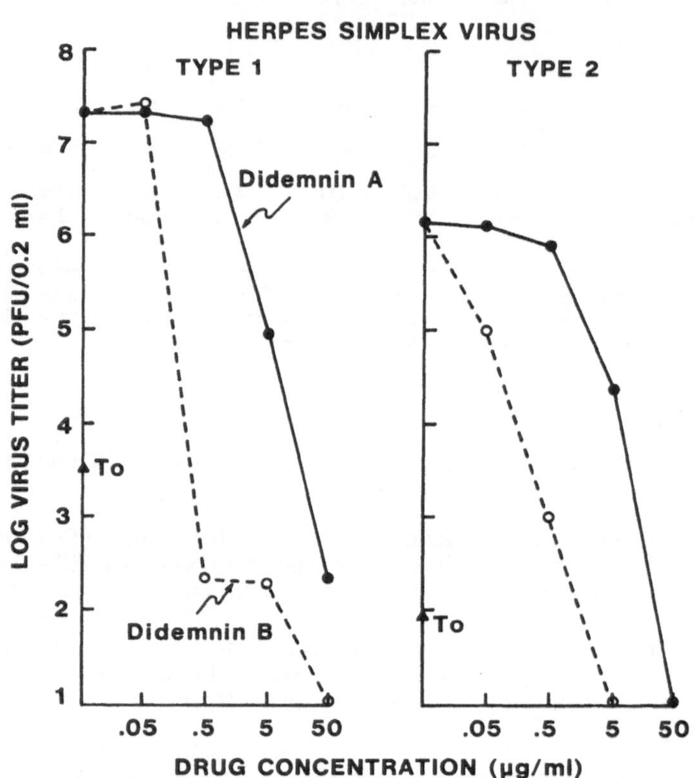

Fig. 1. Effect of didemnins A and B on *Herpes simplex* virus, types 1 and 2, from Vero cell cultures infected with HSV-1 or HSV-2 and treated with didemnins A or B at time zero. After 24 h, the virus yields in the infected cell cultures were determined. Concentrations of didemnins A and B greater than 50 and 12 µg/mL, respectively, were cytotoxic to Vero cells.

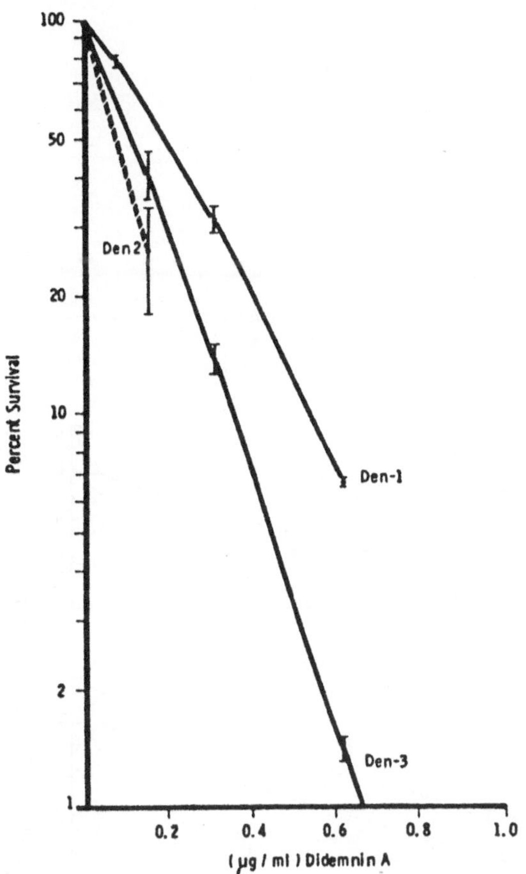

Fig. 2. Effect of didemnin A on the plaque-forming ability of D-1 (Jamaica), D-2 (PR-1129), and D-3 (PR-38) Dengue viruses.

The didemnins are as active against RNA viruses as against DNA viruses. In Figure 3 we see at least a four-log viral titer reduction at 0.5 μg/mL against Coxsackie A-21 virus, equine rhinovirus, and parainfluenza virus.[24,31] More interesting RNA viruses were studied by Dr. Peter G. Canonico at the U.S. Army Medical Center--Rift Valley fever, Venezuelan equine encephalitis, Pichinde virus, yellow fever virus, sandfly fever virus[32] (Figure 4). Most of these are totally eliminated in vitro at approximately 0.1 μg/mL, but Pichinde virus requires a higher concentration. Ribavirin is also effective against these viruses in vitro, but in general about 200 μg/mL is needed to achieve the same effect as 0.1 μg/mL of didemnin B.

In vivo studies with didemnin B have been carried out against HSV-2 and Rift Valley fever. In the Upjohn HSV-2 study (Figure 5), most untreated mice died within 12 days, but with an appropriate amount of didemnin B roughly 80% of the mice survived.[31,33] In the Canonico Rift Valley fever study, the virus killed all the mice within seven days, while 90% of them survived over 21 days with didemnin B (Table 6); at four times the effective dose, however, all the mice were dead within five days. Thus, the didemnins are toxic, and one would find great difficulty in treating most viral diseases with them.

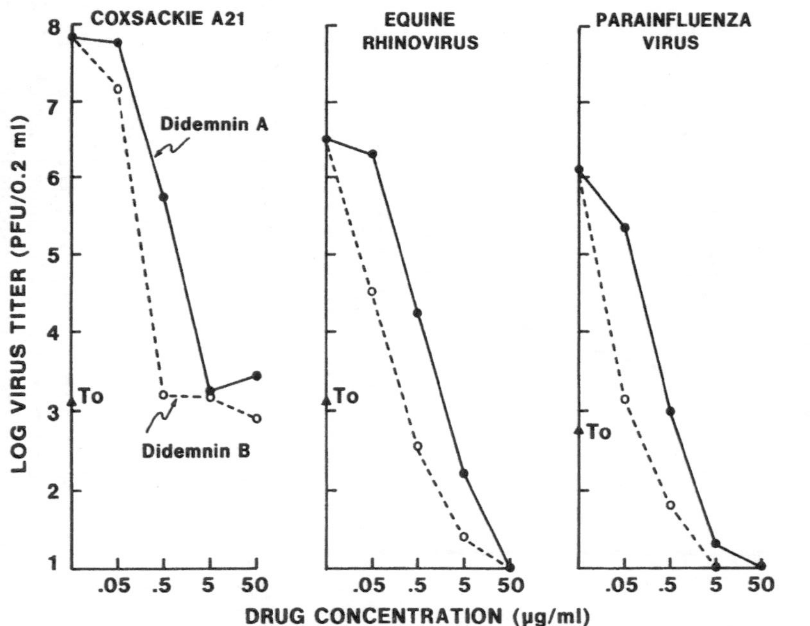

Fig. 3. Effect of didemnins A and B on RNA virus yields from infected HeLa cell (Coxsackie A-21 and equine rhinovirus) or Hep-2 (parainfluenza-3 virus) cultures. The didemnins were added to infected cultures at time zero. Virus yields were determined in the infected cultures 18 h after infection. Didemnins A and B were cytotoxic to non-infected host cells at concentrations greater than 25 and 1.5 μg/mL, respectively.

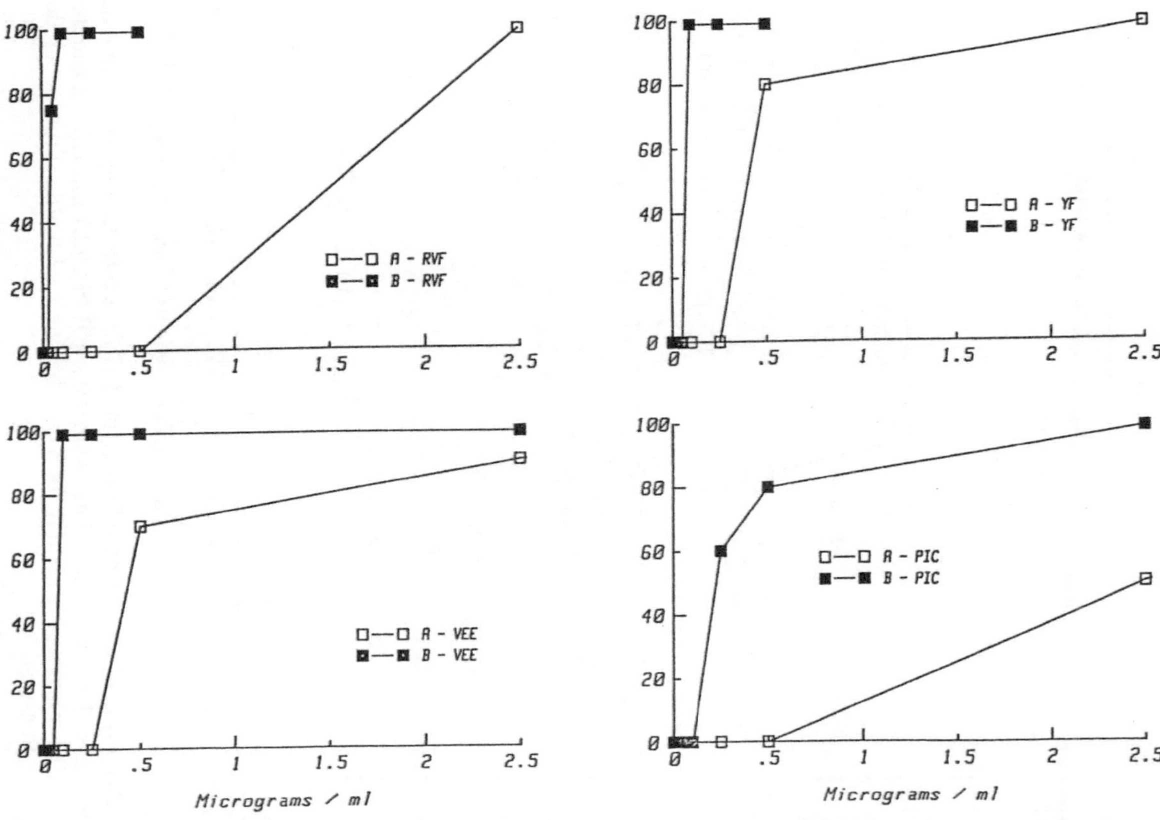

Fig. 4. Viral plaque reduction by didemnins A and B. RVF = Rift Valley fever virus; YF = yellow fever virus; VEE = Venezuelan equine encephalomyelitis; PIC = Pichinde virus.

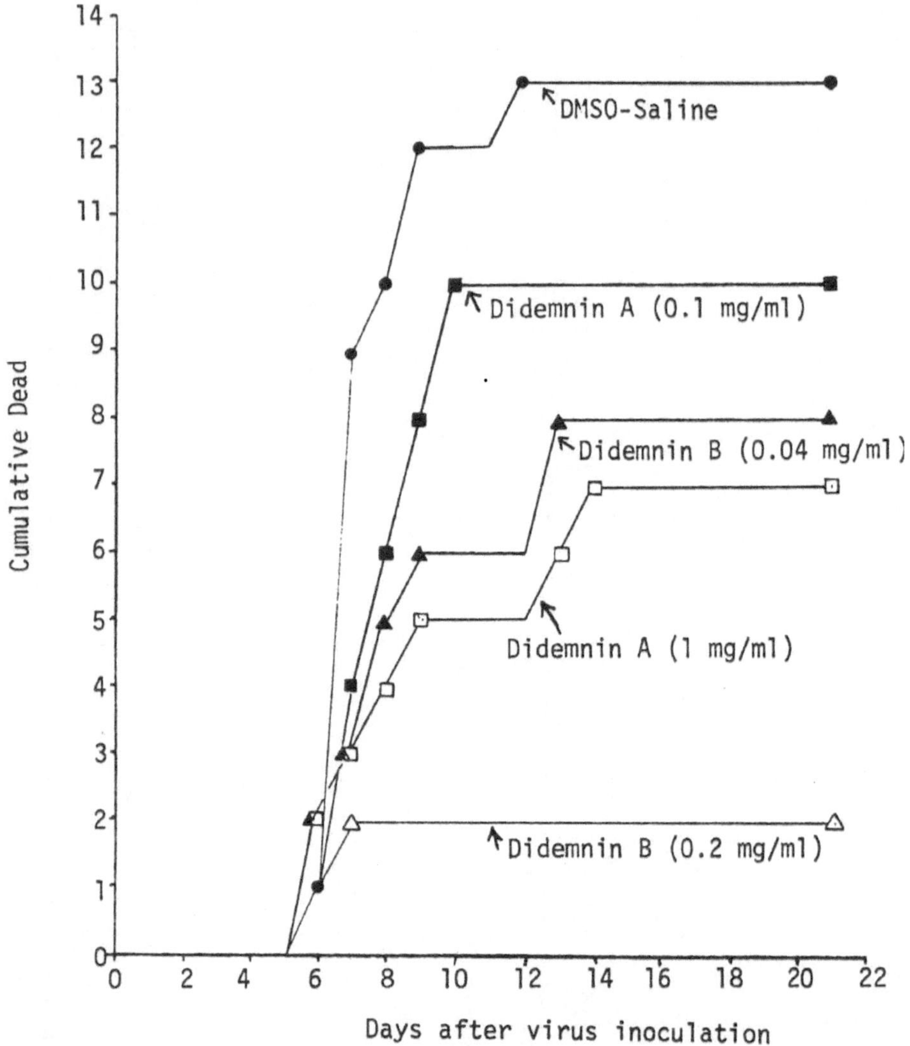

Fig. 5. Protection of female mice from genital HSV-2 infection by IVAG treatment with didemnins. Mice were inoculated with 9.0 x 10^4 PFU/0.1 mL HSV-2 and treated IVAG with 0.1 mL drug 3X per day for 3 days beginning 1 h after virus inoculation. 14 mice in each group.

Table 6. Survival of Rift Valley Fever Virus-Infected Mice Treated wtih Didemnin B[a]

Drug dosage (mg/kg/day)	% Survival by day post-infection						
	1	3	5	7	10	14	21
Virus inoculated							
None (virus control)[b]	100	90	10	0	0	0	0
0.1	100	100	30	10	10	10	10
0.2	100	100	100	50	50	40	40
0.25	100	100	100	90	90	90	90
1.0	100	10	0	0	0	0	0
Uninoculated (drug toxicity controls)							
0.2	100	100	100	100	100	100	100
0.25	100	90	90	90	90	90	80
1.0	100	10	0	0	0	0	0

[a]Didemnin B was given daily beginning on day 1 for 5 consecutive days. [b]Ten mice per group were inoculated s.c. with 250-350 pfu of RVF virus.

Table 7. Antiviral Assay Results

Compound	Dose (μg/well)	VSV		HSV-1		A59	
		Cyt	Av	Cyt	Av	Cyt	Av
Thyrsiferol	10	16		16		0	+
	5	16		16		0	–
	1	16		16		0	–
	0.5	14	+++	14	+++		
	0.1	0	+	0	+++		
	0.01	0	–	0	–		
Thyrsiferol acetate	10	16		16		0	+
	5	16		16		0	–
	1	16		16		0	–
	0.5	16		16			
	0.2	14	++	14	+++		
	0.01	0	–	0	++		
Venustatriol	10	16		16		0	+/–
	5	16		16		0	–
	2	16		10	+++	0	–
	0.5	0	++	0	+++	0	
	0.2	0	+	0	++		
	0.01	0	–				

A number of other antiviral studies have been carried out with didemnin B. Didemnin B is ineffective against *Herpes encephalitis*,[31] so it apparently does not pass the blood-brain barrier. Didemnin B is also inactive against Semliki Forest virus[33] and rabies (an in vivo test by F. Bussereau et al., Institut Pasteur) and, most disappointing of all, inactive against HIV (E. N. Kraiselburd; M. Matsukura, National Cancer Institute).

Other Antiviral Marine-Derived Compounds

Other research groups have also found antiviral agents in marine sources. A particularly active compound is mycalamide A, obtained from a *Mycale* sponge by the Munro-Blunt group at the University of Canterbury in New Zealand.[34] A closely related compound, onnamide, has been isolated from an Okinawan sponge by the Higa group.[35] Both are related to pederin, which is isolated from a beetle.[36] In the case of the mycalamides, both in vivo and in vitro assays have been carried out. In an in vivo test against A59 corona virus, all control mice were dead after eight days, but four mice out of eight survived for 14 days when treated with a crude *Mycale* extract. With the pure compounds, only in vitro assays on HSV-1 and polio virus, type 1, have been performed, in which the mycalamides showed activity at 5 ng/disk, comparable to that of the eudistomins. Onnamide was reported to have potent antiviral activity.

Also of interest is thyrsiferol, a terpene isolated from a New Zealand red alga by the Munro-Blunt group.[37] Thyrsiferol, its acetate, and the related venustatriol have good antiviral activity down to 10-100 ng or so against HSV-1, somewhat lower activity against VSV, and still lower activity against A59 (Table 7). Another quite active compound, spongiadiol, was isolated by Kohmoto et al. at SeaPharm/Harbor Branch Oceanographic Institution.[39] It was active against HSV-1 at concentrations as low as 250 ng/mL. The antiviral activity of misakinolide, a macrolide isolated by the Higa group from an Okinawan sponge, is more modest, *ca.* 4 µg/well.[38]

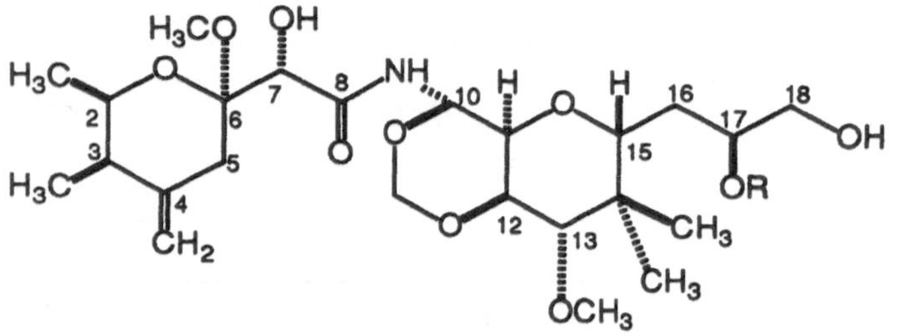

Mycalamide A: R = H
B: R = CH_3

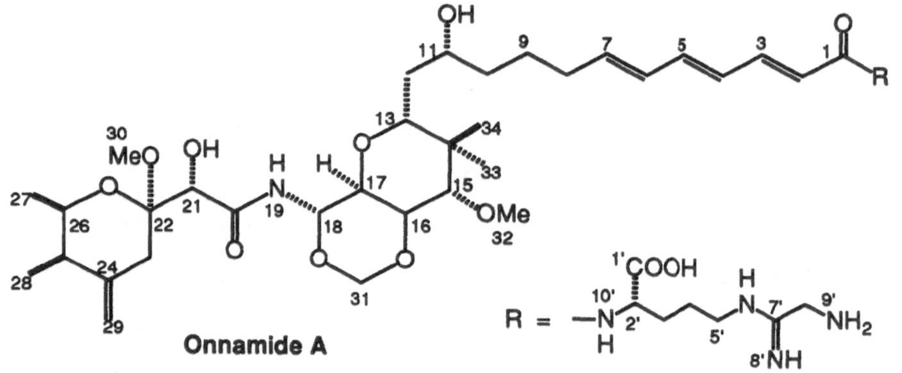

Onnamide A

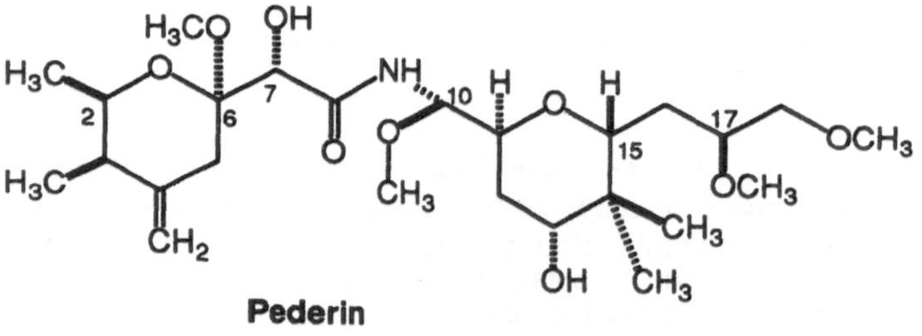

Pederin

Thyrsiferol: 18*S*, 19*R*: R = H
Thyrsiferol acetate: 18*S*, 19*R*: R = Ac
Venustatriol: 18*R*, 19*S*: R = H

Misakinolide

Spongiadiol: $R_1 + R_2 = O$, $R_3 = H$, $R_4 = OH$

R = R' = Ac
R = Ac, R' = i–C_3H_7CO—
R = Ac, R' = C_2H_5CO—
R = H, R' = Ac
R = Ac, R' = H

The Lucibufagins

Antiviral Compounds from Plants and Insects

From the discussion so far it is clear that many different kinds of compounds are to be found in the marine area and that members of quite different structural classes have antiviral activity. About two or three years ago, we began investigating terrestrial plants and insects as other possible sources of antiviral agents. Insects had not previously been studied for their antiviral activity, but of those selected for study by Dr. George R. Wilson in our laboratory approximately 30% have yielded antiviral extracts.[6] A *Photinus* species (firefly) gave very strong antiviral activity in our standard in vitro assay. The compounds isolated from it, the lucibufagins,[40] had been isolated by Dr. Jerrold Meinwald at Cornell[41] but had not been tested for antiviral activity. The lucibufagins are steroids, and the most active inhibit HSV and VSV at a 20-ng level (Table 8). This is about as active as the most potent marine-derived compounds, and we are currently looking at a number of insects, hoping to find other compounds with similar activities.

Table 8. Antiviral Activities of the Lucibufagins

Compound	Dose per well	VSV[a,b]	HSV-1[b,c]
Lucibufagin[d]			
Ac2	780 ng	++	+++
	20 ng	+	+
Ac-Isobutyryl	300 ng	++	++
	30 ng	+	+
Ac-Propionyl	100 ng	+	++
	30 ng	+	+
Mono-Ac	100 ng	+	++
Ribavirin	100 μg	+++	
	10 μg	++	
Ara A	100 μg		+++
	10 μg		+
Acyclovir	10 μg		+++
	100 ng		++

[a]*Vesicular stomatitis* virus. [b]+, Reduced number of plaques, plaque diameter markedly reduced, visible with unaided eye throughout the well; ++, plaques visible with the unaided eye only at the edge of the well, others detected by microscope; +++, no plaques detected by microscope. [c]*Herpes simplex* virus, type 1. [d]No cytotoxicity observed at these doses.

The Supply Problem

A perennial question in discussions of marine natural products as sources of potential medicinal agents is one of supply, and some of the possibilities shown in Table 9 will be discussed here. In a few cases there is enough material in the marine environment to provide for at least Phase II clinical trials. For didemnin B, since it is so active, only 50 grams was required for the Phase II clinical trial, and that was obtained from roughly 1000 lbs of *T. solidum*. The National Cancer Institute has embarked on a clinical study of bryostatin 1 as an antitumor agent[42] that required the collection of 10,000 gallons of the producing bryozoan, *Bugula neritina*; fortunately, *B. neritina* is abundant. Manoalide was in Phase I clinical trial as an antiinflammatory agent,[43] and that compound was isolated from a sponge in large enough quantity. One of the early interests in marine pharmaceuticals centered on a prostaglandin isolated from a particular gorgonian, *Pseudoplexaura homomalla*, where the compound constituted *ca.* 10% of the soft coral's weight.[44]

Table 9. Potential Sources of Marine-Derived Drugs

Natural Sources	Chemical Synthesis
Didemnin B (*Trididemnum solidum*)	Ptilocaulin
Bryostatin (*Bugula neritina*)	Eudistomins
Manoalide (*Luffariella variabilis*)	Didemnins
Prostaglandins	Mycalamide
	Manoalide
	Pseudopterosin B
	Palytoxin
	Halichondrin B (?)
Cultivation	**Genetic Engineering**
Algae	
Fish	
Cyanobacteria	
Dinoflagellates (Saxitoxin)	
Marine Bacteria	
Symbiosis	
sponge bacterial symbionts	
didemnin (?)	

Some marine organisms can be cultivated--algae, cyanobacteria, marine bacteria, and dinoflagellates. In this connection we can note that several compounds from sponges have been shown to be produced by symbiotic bacteria or other microorganisms. For that matter there is a question as to the origin of the didemnins because *T. solidum* grows symbiotically with cyanobacteria,[45,46] and it is conceivable that at least parts of the didemnins are produced by the blue-green alga.

Ultimately, one might hope to transfer the genes which are involved in producing a desired pharmaceutical agent to an organism like *E. coli*, but that will require much basic research on invertebrates.

At the moment, however, chemical synthesis provides the most general route to larger quantities. Eudistomins, including the most active ones, have now been synthesized by groups in four different countries.[12-15,47-49] Didemnins, similarly, have been synthesized, first at Illinois,[22] but within the next two years by another four groups.[50-53] Mycalamide has been synthesized in Dr. Yoshito Kishi's group at Harvard.[54] We can rest assured that organic chemists, if required, can synthesize even the most complicated compounds.

Conclusion. I should like to close with a comment and a prediction. It is possible that we shall never see a marine natural product become a profitable pharmaceutical agent, though that hope certainly still exits. Regardless, however, we can confidently predict that marine natural products will serve as very useful models for new types of structures with particular activities. In this connection spongouridine and spongothymidine, the original arabinoside nucleosides, were reported from a sponge by Dr. Werner Bergmann around 1951,[55] but it was over 10 years before cytosine arabinoside came on the market as an antitumor agent[56] and another 10 years after that, roughly 20 years after the isolation of the sponge compounds, before adenine arabinoside[57] was approved as an antiviral agent. Our confident expectation for the future is that we shall continue to find compounds with promising activity and with interesting structures to at least serve as models for synthetic pharmaceutical agents.

Acknowledgment. Our work has been supported generously through the years by the National Institute of Allergy and Infectious Diseases. Mass spectrometry, which played a major role in our structure determinations, was supported in part by the National Institute of General Medical Sciences, and the *R/V Alpha Helix* was provided by the National Science Foundation.

References

1. K. L. Rinehart, Jr., P. D. Shaw, L. S. Shield, J. B. Gloer, G. C. Harbour, M. E. S. Koker, D. Samain, R. E. Schwartz, A. A. Tymiak, D. L. Weller, G. T. Carter, M. H. G. Munro, R. G. Hughes, Jr., H. E. Renis, E. B. Swynenberg, D. A. Stringfellow, J. J. Vavra, J. H. Coats, G. E. Zurenko, S. L. Kuentzel, L. H. Li, G. J. Bakus, R. C. Brusca, L. L. Craft, D. N. Young, and J. L. Connor, Marine natural products as sources of antiviral, antimicrobial, and antineoplastic agents, *Pure Appl. Chem.* 53:795-817 (1981).
2. A. C. Schroeder, R. G. Hughes, Jr., and A. Bloch, Synthesis and biological effects of acyclic pyrimidine nucleoside analogues, *J. Med. Chem.* 24:1078-1083 (1981).
3. K. L. Rinehart, Screening to detect biological activity, *in* "Biomedical Importance of Marine Organisms," Memoirs of the California Academy of Sciences Number 13, D. G. Fautin, Ed., California Academy of Sciences, San Francisco (1988); pp 13-22.
4. T. G. Holt, The isolation and structural characterization of the ecteinascidins, Ph.D. Dissertation, University of Illinois, Urbana, 1986; *Chem. Abstr.* 106:193149u (1987); *Diss. Abstr. Int. B* 47:3771-3772 (1987).
5. K. L. Rinehart and L. S. Shield, Novel bioactive natural products from marine organisms, *in* "Topics in Pharmaceutical Sciences 1989," D. D. Breimer, D. J. A. Crommelin, and K. K. Midha, Eds., Amsterdam Medical Press B.V., Noordwijk, Netherlands (1989); pp 613-626.
6. K. L. Rinehart, T. G. Holt, N. L. Fregeau, P. A. Keifer, G. R. Wilson, T. J. Perun, Jr., R. Sakai, A. G. Thompson, J. G. Stroh, L. S. Shield, D. S. Seigler, L. H. Li, D. G. Martin, C. J. P. Grimmelikhuijzen, and G. Gäde, Bioactive compounds from aquatic and terrestrial sources, *J. Nat. Prod.* 53:771-792 (1990).
7. P. A. Keifer, R. E. Schwartz, M. E. S. Koker, R. G. Hughes, Jr., D. Rittschof, and K. L. Rinehart, Bioactive bromopyrrole metabolites from the Caribbean sponge *Agelas conifera*, *J. Org. Chem.* 56:000-000 (1991).
8. K. L. Rinehart, Jr., Bioactive metabolites from the Caribbean sponge *Agelas coniferin*, U.S. Patent No. 4,737,510, April 12, 1988; *Chem. Abstr.* 109:216002u (1988).
9. R. P. Walker, D. J. Faulkner, D. Van Engen, and J. Clardy, Sceptrin, an antimicrobial agent from the sponge *Agelas sceptrum*, *J. Am. Chem. Soc.* 103:6772-6773 (1981).
10. K. L. Rinehart, Jr., J. Kobayashi, G. C. Harbour, J. Gilmore, M. Mascal, T. G. Holt, L. S. Shield, and F. Lafargue, Eudistomins A-Q, β-carbolines from the antiviral Caribbean tunicate *Eudistoma olivaceum*, *J. Am. Chem. Soc.* 109:3378-3387 (1987).
11. K. L. Rinehart, Jr., G. C. Harbour, and J. Kobayashi, Antiviral eudistomins from a marine tunicate, U.S. Patent No. 4,631,149, Dec. 23, 1986; *Chem. Abstr.* 102:226023w (1985).
12. M. Nakagawa, J.-J. Liu, and T. Hino, Total synthesis of (-)-eudistomin L and (-)-debromoeudistomin L, *J. Am. Chem. Soc.* 111:2721-2722 (1989).
13. P. H. H. Hermkens, J. H. v. Maarseveen, H. C. J. Ottenheijm, C. G. Kruse, and H. W. Scheeren, Intramolecular Pictet-Spengler reaction of *N*-Alkoxytryptamines. 3. Stereoselective synthesis of (-)-debromoeudistomin L and (-)-*O*-methyldebromoeudistomin E and their stereoisomers, *J. Org. Chem.* 55:3998-4006 (1990).
14. I. W. J. Still and J. R. Strautmanis, "Synthesis of N(10)-acetyleudistomin L, *Tetrahedron Lett.* 30:1041-1044 (1989).
15. M. P. Kirkup, B. B. Shankar, S. McCombie, A. K. Ganguly, and A. T. McPhail, A concise route to the oxathiazepine containing eudistomin skeleton and some carba-analogs, *Tetrahedron Lett.* 30:6809-6812 (1989).
16. K. L. Rinehart, Jr., J. B. Gloer, J. C. Cook, Jr., S. A. Mizsak, and T. A. Scahill, Structures of the didemnins, antiviral and cytotoxic depsipeptides from a Caribbean tunicate, *J. Am. Chem. Soc.* 103:1857-1859 (1981).
17. K. L. Rinehart, Jr., Didemnins A, B, C, and derivatives thereof, as antiviral agents, U.S. Patent No. 4,493,796, Jan. 15, 1985; *Chem. Abstr.* 103:76241v (1985).
18. R. Sakai and K. L. Rinehart, Didemnin X and Y: new cytotoxic cyclic depsipeptides from the tunicate *Trididemnum solidum*, 197th ACS National Meeting, Dallas, TX, April 9-14, 1989, Abstract ORGN 171.
19. K. L. Rinehart, R. Sakai, and J. G. Stroh, Novel cytotoxic cyclic depsipeptides from the tunicate *Trididemnum solidum*, U.S. Patent Appl. Ser. No. 335,903, Apr. 10, 1989.
20. K. L. Rinehart, Novel anti-viral and cytotoxic agents, British Patent Appl. #8922026.3, Sep. 29, 1989.

21. K. L. Rinehart, V. Kishore, K. C. Bible, R. Sakai, D. W. Sullins, and K.-M. Li, Didemnins and tunichlorin: novel natural products from the marine tunicate *Trididemnum solidum*, *J. Nat. Prod.* 51:1-21 (1988).
22. K. L. Rinehart, V. Kishore, S. Nagarajan, R. J. Lake, J. B. Gloer, F. A. Bozich, K.-M. Li, R. E. Maleczka, Jr., W. L. Todsen, M. H. G. Munro, D. W. Sullins, and R. Sakai, Total synthesis of didemnins A, B, and C, *J. Am. Chem. Soc.* 109:6846-6848 (1987).
23. K. L. Rinehart, Jr., J. B. Gloer, R. G. Hughes, Jr., H. E. Renis, J. P. McGovren, E. B. Swynenberg, D. A. Stringfellow, S. L. Kuentzel, and L. H. Li, Didemnins: antiviral and antitumor depsipeptides from a Caribbean tunicate, *Science* 212:933-935 (1981).
24. K. L. Rinehart, Jr., J. B. Gloer, G. R. Wilson, R. G. Hughes, Jr., L. H. Li, H. E. Renis, and J. P. McGovren, Antiviral and antitumor compounds from tunicates, *Fed. Proc.* 42:87-90 (1983)
25. K. L. Rinehart, Jr., J. C. Cook, Jr., R. C. Pandey, L. A. Gaudioso, H. Meng, M. L. Moore, J. B. Gloer, G. R. Wilson, R. E. Gutowsky, P. D. Zierath, L. S. Shield, L. H. Li, H. E. Renis, J. P. McGovren, and P. G. Canonico, Biologically active peptides and their mass spectra, *Pure Appl. Chem.* 54:2409-2424 (1982).
26. K. L. Rinehart, Didemnin and its biological properties, *in* "Peptides. Chemistry and Biology," G. R. Marshall, Ed., ESCOM, Leiden, The Netherlands (1988); pp 626-631.
27. H. G. Chun, B. Davies, D. Hoth, M. Suffness, J. Plowman, K. Flora, C. Grieshaber, and B. Leyland-Jones, Didemnin B: the first marine compound entering clinical trials as an antineoplastic agent, *Invest. New Drugs* 4:279-284 (1986).
28. D. W. Montgomery, A. Celniker, and C. F. Zukoski, Didemnin B--an immunosuppressive cyclic peptide that stimulates murine hemagglutinating antibody responses and induces leukocytosis in vivo, *Transplantation* 43:133-139 (1987).
29. D. H. Russell, A. R. Buckley, D. W. Montgomery, N. A. Larson, P. W. Gout, C. T. Beer, C. W. Putnam, C. F. Zukoski, and R. Kibler, Prolactin-dependent mitogenesis in Nb 2 node lymphoma cells: effects of immunosuppressive cyclopeptides, *J. Immunol.* 138:276-284 (1987).
30. E. Maldonado, J. A. Lavergne, and E. Kraiselburd, Didemnin A inhibits the in vitro replication of dengue virus types 1, 2 and 3, *Puerto Rico Health Sci. J.* 1:22-25 (1982).
31. H. E. Renis, B. A. Court, E. E. Eidson, E. B. Swynenberg, J. B. Gloer, and K. L. Rinehart, Jr., Didemnins--antiviral properties of depsipeptides from a marine tunicate, 21st Intersci. Conf. Antimicrob. Agents Chemother., Chicago, IL, Nov. 4-6, 1981, Abstract #189.
32. P. G. Canonico, W. L. Pannier, J. W. Huggins, and K. L. Rienehart, Inhibition of RNA viruses in vitro and in Rift Valley fever-infected mice by didemnins A and B, *Antimicrob. Agents Chemother.* 22:696-697 (1982).
33. S. D. Weed and D. A. Stringfellow, Didemnins A and B. Effectiveness against cutaneous herpes simplex virus in mice, *Antiviral Res.* 3:269-274 (1983).
34. N. B. Perry, J. W. Blunt, M. H. G. Munro, and A. M. Thompson, Antiviral and antitumor agents from a New Zealand sponge, *Mycale* sp. 2. Structures and solution conformations of mycalamides A and B, *J. Org. Chem.* 55:223-227 (1990).
35. S. Sakemi, T. Ichiba, S. Kohmoto, G. Saucy, and T. Higa, Isolation and structure elucidation of onnamide A, a new bioactive metabolite of a marine sponge, *Theonella* sp., *J. Am. Chem. Soc.* 110:4851-4853 (1988).
36. C. Cardani, D. Ghiringhelli, R. Mondelli, and A. Quilico, The structure of pederin, *Tetrahedron Lett.* 1965:2537-2545 (1965).
37. J. W. Blunt, M. P. Hartshorn, T. J. McLennan, M. H. G. Munto, W. T. Robinson, and S. C. Yorke, Thyrsiferol: a squalene-derived metabolite of *Laurencia thyrsifera*, *Tetrahedron Lett.* 1978:69-72 (1978).
38. Y. Kato, N. Fusetani, S. Matsunaga, K. Hashimoto, R. Sakai, T. Higa, and Y. Kashman, Antitumor macrodiolides isolated from a marine sponge *Theonella* sp.: structure revision of misakinolide A, *Tetrahedron Lett.* 28:6225-6228 (1987).
39. S. Kohmoto, O. J. McConnell, A. Wright, and S. Cross, Isospongiadiol, a cytotoxic and antiviral diterpene from a Caribbean deep water marine sponge, *Spongia* sp., *Chem. Lett.* 1687-1690 (1987).
40. G. R. Wilson and K. L. Rinehart, Antiviral compositions derived from fireflies and their methods of use, U.S. Patent No. 4,847, 246, Jul. 11, 1989; *Chem. Abstr.* 112:48775q (1990).
41. J. Meinwald, D. F. Wiemer, and T. Eisner, Lucibufagins. 2. Esters of 12-oxo-2β,5β,11α-trihydroxybufalin, the major defensive steroids of the firefly *Photinus pyralis* (Coleoptera: Lampyridae), *J. Am. Chem. Soc.* 101:3055-3060 (1979).

42. G. R. Pettit, J. E. Leet, C. L. Herald, Y. Kamano, F. E. Boettner, L. Baczynskyj, and R. A. Neiman, Isolation and structure of bryostatins 12 and 13, *J. Org. Chem.* 52:2854-2860 (1987).
43. R. S. Jacobs, P. Culver, R. Langdon, T. O'Brien, and S. White, Some pharmacological observations on marine natural products, *Tetrahedron* 41:981-984 (1985).
44. A. J. Weinheimer and R. L. Spraggins, "The occurrence of two new prostaglandin derivatives (15-epi-PGA_2 and its acetate, methyl ester) in the gorgonian *Plexaura homomalla*. Chemistry of coelenterates. XV., *Tetrahedron Lett.* 1969:5185-5188 (1969).
45. F. Lafargue and G. Duclaux, Premier exemple, en Atlantique tropical, d'une association symbiotique entre une ascidie didemnidae et une cyanophycée chroococcale: *Trididemnum cyanophorum* nov. sp. et *Synechocystis trididemni* nov. sp., *Ann. Inst. océanogr.* 55:163-184 (1979).
46. J. Sybesma, F. C. van Duyl, and R. P. M. Bak, The ecology of the tropical compound ascidian *Trididemnum solidum*. III. Symbiotic association with unicellular algae, *Mar. Ecol. Prog. Ser.* 6:53-59 (1981).
47. T. Hino, Z. Lai, H. Seki, R. Hara, T. Kuramochi, and M. Nakagawa, Synthesis of eudistomins H, I, and P, β-carboline derivatives from *Eudistoma olivaceum* with antiviral and antimicrobial activity, *Chem. Pharm. Bull.* 37:2596-2600 (1989).
48. B. C. VanWagenen and J. H. Cardellina II, Short, efficient syntheses of the antibiotics eudistomins I and T, *Tetrahedron Lett.* 30:3605-3608 (1989).
49. H. H. Wasserman and T. A. Kelly, The chemistry of vicinal tricarbonyl compounds. Short syntheses of eudistomins T, I, and M, *Tetrahedron Lett.* 30:7117-7120 (1989).
50. Y. Hamada, Y. Kondo, M. Shibata, and T. Shioiri, Efficient total synthesis of didemnins A and B, *J. Am. Chem. Soc.* 111:669-673 (1989).
51. U. Schmidt, M. Kroner, and H. Griesser, Total synthesis of the didemnins - 2. Synthesis of didemnin A, B, C and Prolyldidemnin A, *Tetrahedron Lett.* 29:4407-4408 (1988).
52. W.-R. Li, W. R. Ewing, B. D. Harris, and M. M. Joullié, Total synthesis and structural investigations of didemnins A, B, and C, *J. Am. Chem. Soc.* 112:7659-7672 (1990).
53. P. Jouin, J. Poncet, M.-N. Dufour, A. Pantaloni, and B. Castro, Synthesis of the cyclodepsipeptide nordidemnin B, a cytotoxic minor product isolated from the sea tunicate *Trididemnum cyanophorum*, *J. Org. Chem.* 54:617-627 (1989).
54. C. Y. Hong and Y. Kishi, Total synthesis of mycalamides A and B, *J. Org. Chem.* 55:4242-4245 (1990).
55. W. Bergmann and R. J. Feeney, Contributions to the study of marine products. XXXII. The nucleosides of sponges. I. *J. Org. Chem.* 16:981-987 (1951).
56. J. S. Evans, E. A. Musser, G. D. Mengel, K. R. Forsblad, and J. H. Hunter, Antitumor activity of 1-β-D-arabinofuranosylcytosine hydrochloride, *Proc. Soc. Exp. Biol. Med.* 106:350-353 (1961).
57. D. Pavan-Langston, R. A. Buchanan, and C. A. Alford, Jr., "Adenosine Arabinoside: An Antiviral Agent," Raven Press, New York (1975).

VIRUS RECEPTORS: THE ACHILLES' HEEL OF HUMAN RHINOVIRUSES

Richard J. Colonno

Department of Virus and Cell Biology
Merck Sharp & Dohme Research Laboratories
West Point, PA 19486

INTRODUCTION

Human Rhinoviruses (HRVs), members of the Picornaviridae, are the major causative agents of the common cold in humans (Gwaltney, Jr., 1982). There are currently 102 recognized serotypes that have been isolated and shown to be antigenically distinct (Hamparian et al., 1987). Similar to other picornaviruses, HRVs are non-enveloped viruses that contain four structural proteins, designated VP1, VP2, VP3, and VP4, which form a protein capsid with icosahedral symmetry. Within the viral capsid lies a single-stranded genome RNA which serves as a monocistronic mRNA for the synthesis of all 11 structural and non-structural proteins of the virus. Upon entry into a cell, the RNA genome is translated into a large polyprotein which is subsequently cleaved by two viral proteases encoded within the polyprotein (Palmenberg, 1987). The genome RNA alone is sufficient to initiate a viral infection since transfection of cells with HRV genome RNA results in the production of infectious progeny virus (Mizutani and Colonno, 1985).

HRVs have been divided into 3 receptor families by competition binding and cell protection studies. Ninety-one of the 102 HRV serotypes and at least 3 Coxsackievirus A serotypes utilize a single cellular receptor and are designated the "major group", while 10 (HRV-1A, -1B, -2, -29, -30, -31, -44, -47, -49, -62) of the remaining 11 serotypes compete for a second receptor and comprise the "minor group" (Abraham and Colonno, 1984; Colonno et al., 1986; Uncapher et al., 1991). One serotype, HRV-87, appears to utilize neither the major nor minor group receptor and may represent a third receptor group. HRV-87, which is clearly not a major group virus, displays cell binding tropisms similar but distinct from the minor group viruses, yet is unlike minor group viruses in requiring sialic acid for attachment (Uncapher et al., 1991).

CHARACTERIZATION AND IDENTITY OF THE MAJOR GROUP RECEPTOR

Characterization of the HRV major group receptor was made possible by the isolation of a murine monoclonal antibody, designated MAb 1A6, which inhibited the binding of major group viruses (Colonno et al., 1986). The major group viruses exhibit

Innovations in Antiviral Development and the Detection of Virus Infection
Edited by T. Block *et al.*, Plenum Press, New York, 1992

an absolute requirement for this receptor since blocking attachment with MAb 1A6 was cytoprotective despite high titer viral challenge (Colonno et al., 1986). In addition, MAb 1Ab was capable of displacing virions already bound to viral receptors. HRV attachment appears to be a reversible process since virus released from cellular receptors by MAb 1A6 or other viruses is completely intact and infectious (Abraham and Colonno, 1988).

Using MAb 1A6 immunoaffinity chromatography, a 90 kDa surface glycoprotein with a pI of 4.2 was isolated from HeLa cells (Tomassini and Colonno, 1986; Tomassini et al., 1989a). Carbohydrates accounted for 30% of the molecular mass of the protein and 7 N-linked glycosylation sites were predicted based on partial digestion with N-glycanase (Tomassini et al., 1989a). Digestion with neuraminidase caused a downward mobility shift of 10 kDa on SDS polyacrylamide gels, revealing the presence of sialic acid in the oligosaccharide component of the receptor protein (Tomassini et al., 1989a). Further digestion of desialyated receptor with beta-galactosidase resulted in an additional downward mobility shift of 2 kDa on SDS polyacrylamide gels, indicating the successive linkage of beta-galactose to sialic acid in the oligosaccharide chain.

Using degenerate oligonucleotides, representing deduced peptide sequence from receptor peptides as probes, four overlapping clones were identified and joined together to generate a single clone containing a 3 kb insert (Tomassini et al., 1989b). The full-length clone had a single large reading frame initiating at nucleotide 72 that encoded 532 amino acids, and contained a 1333 nucleotide 3' noncoding region. *In vitro* translation of RNA transcribed from the cDNA clone resulted in a single polypeptide of 55 kDa that is in close agreement to the 54-60 kDa size found for deglycosylated receptor protein (Tomassini et al., 1989a; Tomassini et al., 1989b). DNA sequencing identified the major group receptor as the intercellular adhesion molecule-1 (ICAM-1) (Tomassini et al., 1989b). The identity of ICAM-1 as the major HRV group receptor has also been reported by other laboratories (Greve et al., 1989; Staunton et al., 1989). In addition to the virtually identical sequence homology, both the major group HRV receptor and ICAM-1 receptor proteins have equivalent mass, tissue distribution, and carbohydrate moieties (Colonno et al., 1986; Dustin et al., 1988; Staunton et al., 1988; Tomassini et al., 1989a). Confirmation that the ICAM-1 receptor was the functional receptor for the major group of HRVs was obtained by generating a stable Vero cell line and showing that the expression of ICAM-1 on the surface of Vero cells enabled HRV major group binding and infection (Colonno et al., 1990).

ICAM-1

Apart from serving as the cellular receptor for major group HRVs, ICAM-1 functions as the cell surface ligand for lymphocyte function-associated antigen 1 (LFA-1). The interaction between ICAM-1 and LFA-1 plays an important role in several immunological and inflammatory functions including adherence of leukocytes to endothelial cells, fibroblasts, and epithelial cells, responses by T-helper and beta-lymphocytes, neutrophil recruitment, natural killing, and antibody-dependent cytotoxicity mediated by monocytes and granulocytes, (Springer et al., 1987; Dustin et al., 1988; Makgoba et al., 1988; Kishimoto et al., 1989; Osborn, 1990). In contrast to LFA-1, ICAM-1 is expressed on a diversity of human cells and its strong inducibility by cytokines plays an important role in regulating inflammatory responses (Springer, 1990). The ICAM-1 receptor is a member of the immunoglobulin supergene family and is predicted to have 5 homologous immunoglobulin-like domains defined by amino acids 1-88, 89-185, 186-284, 285-385, and 386-453 (Stauton et al., 1988). In addition, ICAM-1 is closely related to 2 adhesion proteins of the adult nervous system, namely, neural cell adhesion molecule and myelin-associated glycoprotein (Simmons et al., 1988; Staunton et al., 1988).

The significance of HRVs utilizing a receptor having immunological and inflammatory functions is unclear. HRV infection appears to involve only a limited number of cells in the nasal epithelium and the clinical symptomatology associated with a cold may actually result from the release of inflammatory mediators such as kinins (Turner et al., 1984; Naclerio et al., 1988). Several clinical studies have already demonstrated that HRV infection induces cytokines, such as interleukin-2 and gamma-interferon, causes the release of kinins, and elevates the serum level of thymosins (α1 and β4) involved in modulation of the T cell lymphoproliferative response (Hsia et al., 1989; Proud et al., 1990). It is tempting to speculate that the interaction between HRVs and ICAM-1 somehow plays an important role in the initiation of the systemic immune response observed during common cold infections.

HRV BINDING SITE ON ICAM-1

The prediction that ICAM-1 is structurally a member of the immunoglobulin supergene family suggests that ICAM-1 is related to the CD4 receptor and to the immunoglobulin-like protein recently identified as the cellular receptor for poliovirus (Mendelsohn et al., 1989). The CD4 receptor utilized by the human immunodeficiency virus (HIV) is postulated to have 4 domains, while the poliovirus receptor is estimated to have 3 homologous immunoglobulin-like domains (Littman and Gettner, 1987; Mendelssohn et al., 1989). Several laboratories have mapped the cellular binding site for HIV gp120 to the N-terminal domain (Clayton et al., 1988; Jameson et al., 1988; Peterson et al., 1988; Arthos et al., 1989). Anti-CD4 MAbs (OKT4A, Leu3A) that block HIV attachment also map to this same region (Jameson et al., 1988; Landau et al., 1988; Peterson et al., 1988) and suggest that MAbs capable of abrogating virus attachment can be utilized to map the site of virus interaction.

Studies were undertaken to map the ICAM-1 binding sites of MAb 1A6 and 2 additional MAbs, 2C2 and 18B9, which showed comparable efficiency in blocking HRV attachment and bind to non-overlapping sites on ICAM-1 (Lineberger et al., 1990). Fragments of the ICAM-1 receptor were generated by *in vitro* transcription and cell-free translation. Microsomal membranes, which are only capable of core glycosylation, were found to be required for proper folding and/or glycosylation since immunoprecipitation experiments showed that only the proteins that had transversed the membranes were immunoprecipitated by the 3 MAbs (Lineberger et al., 1990). Interestingly, this was true for the shortest ICAM-1 fragment (domain 1) which is not predicted to have glycosylation sites, and further suggests that insertion into membranes confer a conformational effect on *in vitro* synthesized ICAM-1.

Experiments utilizing the subset of ICAM-1 polypeptides clearly showed that all 3 of the anti-ICAM-1 MAbs assayed were able to recognize an ICAM-1 fragment containing only the N-terminal 82 amino acid residues (Lineberger et al., 1990). It is interesting to note that this small fragment contains 2 putative SH bonds bridging Cys residues at positions 21 and 25 with Cys residues 65 and 69, respectively (Tomassini et al, 1989b). The 2 putative SH bridges in domain 1 appear to be important for conformational integrity since shorter fragments deleting Cys residues 65 and 69 result in fragments no longer recognized by any of the MAbs (Lineberger et al., 1990).

While virus binding to purified ICAM-1 receptor isolated from HeLa cells can be easily demonstrated, attempts to show specific virus binding with either the full-length or domain 1 *in vitro* fragments were unsuccessful (Lineberger et al., 1990). Interestingly, unglycosylated ICAM-1, purified from HeLa cells treated with tunicamycin to prevent N-linked glycosylation, also failed to bind virus. These virus binding studies suggest that glycosylation may play a role in the generation of a functional receptor protein that is more important for virus binding than MAb binding. The absence of carbohydrates could deter HRV binding directly if it were itself part of the virus binding site, or indirectly by causing an alternative conformation of the native protein that is incapable of efficient interaction with the virion attachment site. In addition, it has also been suggested that the cellular receptor site may be a mutimeric complex composed of five 90 kDa receptor proteins (Tomassini and Colonno, 1986). It is possible that glycosylation and/or additional domains may play a role in the formation of such pentamers. It should be noted that domain 1 of ICAM-1 contains none of the predicted glycosylation sites present in ICAM-1, while neighboring domains 2 and 3 are predicted to account for 6 N-linked glycosylation sites (Tomassini et al., 1989a). In contrast to the ICAM-1/HRV interaction, glycosylation of the CD4 receptor does not appear to play a functional role in HIV binding since no glycosylation sites exist within the first domain of CD4 and digestion of carbohydrate groups did not affect binding (Ibegbu et al., 1989). Recent studies by Staunton et al. (1990), using transient expression of ICAM-1 mutant and chimerics have also localized the primary HRV binding site to domain 1 and suggest that domains 1 and 2 interact conformationally. They also indicated that truncation of the C-terminal domains of ICAM-1 reduces virus binding by altering ICAM-1 accessibility.

VIRION ATTACHMENT SITE OF HUMAN RHINOVIRUSES

Crystallographic studies on the major group of virus HRV-14 led Rossmann et al. (1985) to propose that a deep crevice or canyon found on the surface of the virus may be involved in receptor interaction. A similar surface depression has also been previously described for the sialic acid binding site of the hemagglutinin protein of influenza (Wiley and Skehel, 1987). Placement of virion receptor binding sites in such crevices allows maintenance of a highly conserved surface site among a diversity of antigenically distinct serotypes that excludes the production and/or accessibility of cross-neutralizing immunoglobulins by the infected host. The concept of the HRV canyon as the receptor attachment site is supported by the finding that 91 HRV serotypes utilize a highly-conserved attachment site to which the host immune system fails to generate antibodies capable of cross neutralizing a vast number of serotypes (Colonno et al., 1986; Uncapher et al., 1991) and by the failure of neutralizing anti-idiotypic antisera to MAb 1A6 to recognize HRVs (Colonno, 1987).

The most direct evidence was obtained from recombinant DNA studies in which a series of single amino acid changes were introduced into the VP1 protein of HRV-14 at 4 amino acid residues that reside on the floor of the canyon. Results indicated that changing the Lys 103 to Asn, His 220 to Trp or Ile, or changing Ser 223 to Thr, Ala, or Asn resulted in HRV-14 mutants with reduced binding capabilities (Colonno et al., 1988). Unexpectedly, substitution of Gly for a Pro at residue 155 resulted in a mutant virus that displayed a binding K_D some 9 times greater than the parent HRV-14 from which it was derived (Colonno et al., 1988). The binding phenotypes displayed by the HRV-14 mutants represent conclusive evidence that the canyon floor is involved in receptor interaction. Whether the observed binding affinities were the result of simple charge changes or reflect gross anatomical changes in the floor of the canyon is not known. However, the possibility that these changes could have affected structures outside of the canyon is remote since a rippling effect, involving numerous amino acids within the wall of the canyon, would be needed to significantly alter external structures some 22 angstroms away. Since this region of the canyon is highly conserved among HRV serotypes, poliovirus, and coxsackieviruses A and B, it is highly unlikely that this particular region is involved in determining receptor specificity. Instead, it is more probable that this conserved region interacts with a common structural domain found on several picornaviral receptors and that it is the shape and contour of the canyon entrance which defines the accessibility of a specific receptor protein into the viral canyon (Colonno et al., 1988).

Additional support for the canyon model has come from studies using a series of compounds that interfere with virus uncoating. These small hydrophobic molecules were shown to enter a small pore located at the bottom of the canyon (Badger et al., 1988). In the case of the major group serotype, HRV-14, these drugs cause amino acids located at the floor of the canyon, including the amino acid residues mentioned above, to move upward into the canyon (Badger et al., 1988). Binding studies on HRV-14 following drug treatment show that displacement of the canyon floor interferes with virus attachment and represents further proof that the canyon harbors the virion attachment site (Pevear et al., 1989).

Computer modeling of ICAM-1 domain 1 proposes that approximately half of domain 1 could occupy the HRV canyon and interact with amino acids located at the bottom of the canyon (Giranda et al., 1990). Although highly speculative in nature, this docking of ICAM-1 and its predicted interactions with the virion canyon are supported by the canyon mutagenesis studies above and the mutagenic studies on ICAM-1 (Colonno et al., 1988; Staunton et al., 1990).

IN VIVO RECEPTOR BLOCKADE OF HRV INFECTION

The finding that MAb 1A6 was capable of establishing an effective receptor blockade in cell culture experiments (Colonno et al., 1986), suggested that MAb 1A6 may be useful as an antiviral agent in the prevention or treatment of common colds in humans. Unfortunately, no animal models exist that duplicate the symptoms experience by humans following HRV infection. Since chimpanzees are the only non-human species shown to have ICAM-1 receptors recognized by MAb 1A6 (Colonno et al., 1986; Tomassini and Colonno, 1986), a chimpanzee model was utilized in which subclinical HRV infection of the nasal cavity was measured by the generation of neutralizing antisera. Results of this study showed that a single intranasal administration (0.5 mg/nostril) of MAb 1A6 prior to HRV challenge blocked the subsequent development of neutralizing antibodies in contrast to placebo-treated animals (Colonno et al., 1987).

Encouraged by the results obtained in the chimpanzee study, 2 human double-blind clinical trials were conducted to assess the safety and efficacy of MAb 1A6 in preventing HRV infection in humans. In the first study, 9 doses of 15 μg of MAb 1A6 were applied intranasally over a 65 hr period with no effect on the clinical course of experimental HRV-39 infection (Hayden et al., 1988). Increasing the dose level to 0.1 mg/dose and administering 10 doses of MAb 1A6 over a shorter time period (36 hr) gave more encouraging results. In this study, a significant reduction in peak virus titers were observed, and delays in the onset of colds, nasal symptoms, and nasal mucous production were noted (Hayden et al., 1988). However, no overall reduction in the frequency of colds was observed in the second study. While there are a number of possible explanations for the failure of MAb 1A6 to prevent colds entirely, the most obvious at this time appears to be insufficient dose levels of the antibody. The finding that MAb 1A6 did alter the clinical course of the disease confirms that ICAM-1 is utilized during HRV infection *in vivo* and suggests that an ICAM-1 receptor antagonist may be useful in the prevention of the common cold.

ANTIBODY FRAGMENTS THAT BLOCK HRV ATTACHMENT

Because large amounts of MAb 1A6 may be required to provide protection *in vivo*, and because intranasal administration of intact murine antibodies may be immunogenic, biologically active sub-fragments of MAb 1A6 were expressed using recombinant DNA technology. Full-length cDNA clones of MAb 1A6 heavy and light chains were isolated and individually expressed in *E. coli.* The expressed antibody fragments were subsequently renatured together to generate recombinant Fab fragments that were only 3-fold less active in HRV binding assays than intact MAb 1A6 (Condra et al., 1990).

In an effort to reduce the size and complexity of the expressed protein, single-chain antibodies (SCAbs) were constructed in which the C-terminus of the light chain variable domain was covalently linked to the N-terminus of the heavy chain variable domain using different linker peptides. The first, SCAb1, linked the Glu 124 of the light chain directly to the Glu 1 of the heavy chain. The remaining 2 recombinants, SCAb2 and SCAb3, incorporated the flexible linker sequence Gly-Gly-Gly-Gly-Ser as a monomer, or dimer, respectively, between the Ser 122 of the light chain and the Glu 1 of the heavy chain. In contrast to the Fab fragments, the SCAbs were 14-fold (SCAb2 & SCAb3) and 40-fold (SCAb1) less active in the HRV binding assay compared to intact antibody (Condra et al., 1990).

To ascertain the potential utility of these recombinant immunoglobulin fragments in preventing viral infection, each was used in cell protection studies. Results showed that MAb 1A6 was again the most effective biological molecule, protecting at a concentration as low as 82 ng/ml (0.6 nM). Similar to the virus binding assays, recombinant Fab fragments were less potent but were still cytoprotective down to 740 ng/ml (16 nM). The SCAbs showed lower but significant activity, and were protective at 2.2-6.7 μg/ml (84-250 nM) (Condra et al., 1990).

These studies show that sub-fragments of recombinant immunoglobulin chains can be abundantly expressed in *E. coli.* and efficiently purified and refolded into biologically active molecules. The cell protection experiments described above show that although monovalent antibody fragments are protective against HRV-14 infection, bivalent molecules are significantly more potent. In contrast to MAb 1A6, which remains bound to cellular receptors in cell culture for up to 2 days (Colonno et al., 1989), bound monovalent fragments are easily removed from cell membranes by washing with buffer. This lowered avidity is most likely due to the inability to bind bivalently since bivalently bound $F(ab')_2$ fragments prepared by papain digestion of MAb 1A6 remain stably bound.

FUTURE STRATEGIES FOR THE DESIGN OF THERAPEUTICS AGAINST HRVs

The containment of HRV infections within the nasal cavity during colds enables potential antivirals to be administered topically to limit drug toxicities in treating such a benign disease. However, intranasal administration presents its own problems since drugs are readily removed from the nasal cavity following administration. Inhibition of HRV attachment to cellular receptors appears to be amenable to topical treatment and represents an excellent target for the design of antivirals. HRV interaction with ICAM-1 receptors can be inhibited either be antagonists of ICAM-1, such as MAb 1A6, or of the virion attachment site, such as uncoating drugs and forms of soluble ICAM-1 receptors (Al-Nakib et al., 1990; Marlin et al., 1990).

The advantages of a strategy involving receptor antagonists are that administered drugs would be anchored tightly to ICAM-1 and not be readily cleared from the nose, and that resistant viruses would be unlikely to arise. A potential disadvantage to this approach is the possible side effects of inhibiting normal cell-adhesion function(s) of ICAM-1, although no toxicity has been observed in chimpanzees or human volunteers treated with MAb 1A6 (Hayden et al., 1988). It may also be possible to design antagonists that block virus attachment without interfering with LFA-1 binding, since initial mapping studies indicate that HRVs and LFA-1 bind to overlapping but distinct sites on ICAM-1 (Staunton et al., 1990). In addition, there are also indications that molecules that bind to ICAM-1, such as MAbs, may have other potential therapeutic applications in the treatment of asthma and graft rejection (Wagner et al., 1990; Cosimi et al., 1990). While recombinantly-expressed MAb 1A6 fragments are cytoprotective, additional sequence optimization and enhancement in expression and renaturation levels will need to be achieved before they are deemed feasible. Instead, efforts are continuing to identify small molecules that bind ICAM-1 and mimic the inhibitory activity of MAb 1A6.

The second approach, involving agents that bind to HRVs and interfere with attachment, will require efficient delivery systems to maintain effective levels of drugs over a prolonged period of time since it may not be feasible to administer drugs several times daily. This was evident in human clinical studies on R61837, an inhibitor of HRV binding and uncoating, in which frequent and prolonged prophylactic treatment was necessary to obtain a delay and reduction in cold symptoms (Al-Nakib et al., 1989). Recent studies have also indicated that soluble ICAM-1 fragments are able to bind to virions and block HRV infection in cell culture (Marlin et al., 1990).

Clearly, much work remains to be done before we will be able to fully understand the cascade of events that lead to the symptoms of a common cold and to design strategies and therapeutics to prevent HRV infections.

REFERENCES

Abraham, G., and Colonno, R.J., 1984, Many rhinovirus serotypes share the same cellular receptor, J. Virol., 51:340-345.

Abraham, G., and Colonno, R.J., 1988, Characterization of human rhinoviruses displaced by an anti-receptor monoclonal antibody, J. Virol., 62:2300-2306.

Al-Nakib, W., Higgins, P.G., Barrow, G.I., Tyrrell, D.A.J., Andries, K., Vanden Bussche, G., Taylor, N., and Janssen, P.A.J., 1989, Suppression of colds in human volunteers challenged with rhinovirus by a new synthetic drug (R61837), Antimicrob. Agents and Chemo., 33:522-525.

Arthos, J., Deen, K.C., Chaikin, M.A., Fornwald, J.A., Sathe, G., Sattentau, Q.J., Clapham, P.R., Weiss, R.A., McDougal, J.S., Pietropaolo, C., Axel, R., Truneh, A., Maddon, P.J., and Sweet, R.W., 1989, Identification of the residues in human CD4 critical for the binding of HIV, Cell, 57:469-481.

Badger, J., Minor, I., Kremer, M.J., Oliveira, M.A., Smith, T.J., Griffith, J.P., Guerin, D.M.A., Krishnaswamy, S., Luo, M., Rossmann, M.G., McKinlay, M.A., Diana, G.D., Dutko, F.J., Fancher, M., Rueckert, R.R., and Heinz, B.A., 1988, Structural analysis of a series of antiviral agents complexed with human rhinovirus 14, Proc. Natl. Acad. Sci. USA, 85:3304-3308.

Clayton, L.K., Hussey, R.E., Steinbrich, R., Ramachandran, H., Husain, Y., and Reinherz, E.L., 1988, Substitution of murine for human CD4 residues identifies amino acids critical for HIV-gp120 binding, Nature, 335:363-366.

Colonno, R.J., 1987, Cell surface receptors for picornaviruses, BioEssays, 5:270-274.

Colonno, R.J., Abraham, G., and Tomassini, J.E., 1989, Molecular and biochemical aspects of human rhinovirus attachment to cellular receptors, in: "Molecular Aspects of Picornavirus Infection and Detection", B.L. Semler and E. Ehrenfeld, eds., American Society for Microbiology, Washington.

Colonno, R.J., Callahan, P.L., and Long, W.J., 1986, Isolation of a monoclonal antibody that blocks attachment of the major group of human rhinoviruses, J. Virol., 57:7-12.

Colonno, R.J., Condra, J.H., Mizutani, S., Callahan, P.L., Davies, M.E., and Murcko, M.A., 1988, Evidence for the direct involvement of the rhinovirus canyon in receptor binding, Proc. Natl. Acad. Sci. USA, 85:5449-5453.

Colonno, R.J., LaFemina, R.L., DeWitt, C.M., and Tomassini, J.E., 1990, The major group rhinoviruses utilize the intercellular adhesion molecule 1 ligand as a cellular receptor during infection, in: "New Aspects of Positive-Strand RNA Viruses", M.A. Brinton and F.X. Heinz, eds., American Society for Microbiology, Washington.

Colonno, R.J., Tomassini, J.E., and Callahan, P.L., 1987, Isolation and characterization of a monoclonal antibody which blocks attachment of human rhinoviruses, in: "Positive Strand RNA Viruses", M.A. Brinton and R. Rueckert, eds., Alan R. Liss, Inc., New York.

Condra, J.H., Sardana, V.V., Tomassini, J.E., Schlabach, A.J., Davies, M.E., Lineberger, D.W., Graham, D.J., Gotlib, L., and Colonno, R.J., 1990, Bacterial expression of antibody fragments that block human rhinovirus infection of cultured cells, J Biol. Chem., 265:2292-2295.

Cosimi, A.B., Conti, D., Delmonico, F.L., Preffer, F.I., Wee, S.L., Rothlein, R., Faanes, R., and Colvin, R.B., 1990, In vivo effects of monoclonal antibody to ICAM-1 (CD54) in nonhuman primates with renal allografts, J. Immunol., 144:4604-4612.

Dustin, M.L., Staunton, D.W., and Springer, T.A., 1988, Supergene families meet i n the immune system, Immunol. Today, 9:213-215.

Giranda, V.L., Chapman, M.S., Rossmann, M.G., 1990, Modeling of the human intercellular adhesion molecule-1, the human rhino-virus major group receptor, Proteins: Struct., Funct., and Genetics, 7:227-233.

Greve, J.M., Davis, G., Meyer, A.M., Forte, C.P., Yost, S.C., Marlow, C.W., Kamarck, M.E., and McClelland, A., 1989, The major human rhinovirus receptor is ICAM-1, Cell, 56:839-847.

Gwaltney, J.M., Jr., 1982, Rhinoviruses, in: "Viral Infection of Man: Epidemiology and Control", E.A. Evans, ed., Plenum Publishing Corp., New York.

Hamparian, V.V., Colonno, R.J., Cooney, M.K., Dick, E.C., Gwaltney, Jr., J.M., Hughes, J.H., Jordan, Jr., W.S., Kapikian, A.Z., Mogabgab, W.J., Monto, A., Phillips, C.A., Rueckert, R.R., Schieble, J.H., Stott, E.J., and Tyrrell, D.A.J., 1987, A collaborative report: rhinoviruses--extension of the numbering system from 89 to 100, Virol., 159:191-192.

Hayden, F.G., Gwaltney, Jr., J.M., and Colonno, R.J., 1988, Modification of experimental rhinovirus colds by receptor blockade, Antiviral Res., 9:233-247.

Hsia, J., Sztein, M.B., Naylor, P.H., Simon, G.L., Goldstein, A.L., and Hayden,F.G., 1989, Modulation of thymosin alpha 1 and thymosin beta 4 levels and peripheral blood mononuclear cell subsets during experimental rhinovirus colds, Lymphokine Res., 8:383-391.

Ibegbu, C.C., Kennedy, M.S., Maddon, P.J., Deen, K.C., Hicks, D., Sweet, R.W., and McDougal, J.S., 1989, Structural features of CD4 required for binding to HIV, J. Immunol., 142:2250-2256.

Jameson, B.A., Rao, P.E., Kong, L.I., Hahn, B.H., Shaw, G.M., Hood, L.E., and Kent, S.B.H., 1988, Location and chemical synthesis of a binding site for HIV-1 on the CD4 protein, Science, 240:1335-1339.

Kishimoto, T.K., Larson, R.S., Corbi, A.L., Dustin, M.L., Staunton, D.W., and Springer, T.A., 1989, The leukocyte integrins, in: "Advances in Immunology", F.J. Dixon, ed., Academic Press, Inc., San Diego.

Landau, N.R., Warton, M., and Littman, D.R., 1988, The envelope glycoprotein of the human immunodeficiency virus binds to the immunoglobulin-like domain of CD4, Nature, 334:159-162.

Lineberger, D.W., Graham, D.J., Tomassini, J.E., and Colonno, R.J., 1990, Antibodies that block rhinovirus attachment map to domain 1 of the major group receptor, J. Virol., 64:2582-2587.

Littman, D.R., and Gettner, S.N., 1987, Unusual intron in the immunoglobulin domain of the newly isolated murine CD4 (L3T4) gene, Nature, 325:453-455.

Makgoba, M.W., Sanders, M.E., Ginther Luce, G.E., Gugel, E.A., Dustin, M.L., Springer, T.A., and Shaw, S., 1988, Functional evidence that intercellular adhesion molecule-1 (ICAM-1) is a ligand for LFA-1 dependent adhesion in T cell-mediated cytotoxicity, Eur. J. Immunol., 18:637-640.

Marlin, S.D., Staunton, D.E., Springer, T.A., Stratowa, C., Sommergruber, W., and Merluzzi, V.J., 1990, A soluble form of intercellular adhesion molecule-1 inhibits rhinovirus infection, Nature, 344:70-72.

Mendelsohn, C.L., Wimmer, E., and Racaniello, V., 1989, Cellular receptor for poliovirus: molecular cloning, nucleotide sequence, and expression of a new member of the immunoglobulin superfamily, Cell, 56:855-865.

Mizutani, S., and Colonno, R.J., 1985, In vitro synthesis of an infectious RNA from cDNA clones of human rhinovirus type 14, J. Virol., 56:628-632.

Naclerio, R.M., Proud, D., Lichtenstein, L.M., Kagey-Sobotka, A., Hendley, J.O., Sorrentino, J., and Gwaltney, J.M., 1988, Kinins are generated during experimental rhinovirus colds, J. Infect. Dis., 157:133-142.

Osborn, L., 1990, Leukocyte adhesion to endothelium in inflammation, Cell, 62:3-6.

Palmenberg, A.C., 1987, Picornaviral processing: some new ideas, J. Cell. Biochem., 33:191-198.

Peterson, A., and Seed, B., 1988, Genetic analysis of monoclonal antibody and HIV binding sites on the human lymphocyte antigen CD4, Cell, 54:65-72.

Pevear, D.C., Fancher, M.J., Felock, P.J., Rossmann, M.G., Miller, M.S., Diana, G., Treasurywala, A.M., McKinlay, M.A., and Dutko, F.J., 1989, Conformational change in the floor of the human rhinovirus canyon blocks adsorption of HeLa cell receptors, J.Virol., 63:2002-2007.

Proud, D., Naclerio, R.M., Gwaltney, J.M., and Hendley, J.O., 1990, Kinins are generated in nasal secretions during natural rhinovirus colds, J. Inf. Dis., 161:120-123.

Rossmann, M.G., Arnold, E., Erickson, J.W., Frankenberger, E.A., Griffith, J.P., Hecht, H.J., Johnson, J.E., Kamer, G., Luo, M., Mosser, A.G., Rueckert, R.R., Sherry, B., and Vriend, G., 1985, Structure of a human common cold virus and functional relationship to other picornaviruses, Nature, 317:145-153.

Simmons, D., Makgoba, M.W., and Seed, B., 1988, ICAM, an adhesion ligand of LFA-1, is homologous to the neural cell adhesion molecule NCAM, Nature, 331:624-627.

Sorrentino, J., and Gwaltney, J.M., 1988, Kinins are generated during experimental rhinovirus colds, J.Infect. Dis., 157:133-142.

Springer, T.A., 1990, Adhesion receptors of the immune system, Nature, 346:425-433.

Springer, T.A., Dustin, M.L., Kishimoto, T.K., and Marlin, S.D., 1987, The lymphocyte function-associated LFA-1, CD2, and LFA-3 molecules: Cell adhesion receptors of the immune system, in: "Annual Review of Immunology", W.E. Paul, ed., Anuual Reviews Inc., Palo Alto.

Staunton, D.E., Dustin, M.L., Erickson, H.P., and Springer, T.A., 1990, The arrangement of the immunoglobulin-like domains of ICAM-1 and the binding sites for LFA-1 and rhinovirus, Cell, 61:243-254.

Staunton, D.E., Marlin, S.D., Stratowa, C., Dustin, M.L., and Springer, T.A., 1988, Primary structure of ICAM-1 demonstrates interaction between members of the immunoglobulin and integrin supergene families, Cell, 52:925-933.

Staunton, D.E., Merluzzi, V.J., Rothlein, R., Barton, R., Marlin, S.D., and Springer, T.A., 1989, A cell adhesion molecule, ICAM-1, is the major surface receptor for rhinoviruses, Cell, 56:849-853.

Tomassini, J.E., and Colonno, R.J., 1986. Isolation of a receptor protein involved in attachment of human rhinoviruses, J. Virol., 58:290-295.

Tomassini, J.E., Graham, D., DeWitt, C.M., Lineberger, D.W., Rodkey, J.A., and Colonno, R.J., 1989b, cDNA cloning reveals that the major group rhinovirus receptor on HeLa cells is intercellular adhesion molecule 1, Proc. Natl. Acad. Sci. USA, 86:4907-4911.

Tomassini, J.E., Maxson, T.R., and Colonno, R.J., 1989a, Biochemical characterization of a glycoprotein required for rhinovirus attachment, J. Biol. Chem., 264:1656-1662.

Turner, R.B., Winther, B., Hendley, J.O., Mygind, N., and Gwaltney, Jr., J.M., 1984, Acta Otolaryngol, 413:9-14.

Uncapher, C.R., DeWitt, C.M., and Colonno, R.J., 1991, The major and minor group receptor families contain all but one human rhinovirus serotype, Virol., 180: 814-817.

Wegner, C.D., Gundel, R.H., Reilly, P., Haynes, N., Letts, L.G., and Rothlein, R., 1990, Intercellular adhesion molecule-1 (ICAM-1) in the pathogenesis of asthma, Science, 247:456-459.

Wiley, D.C., and Skehel, J.J., 1987, The structure and function of the hemagglutinin membrane glycoprotein of influenza virus, Annu. Rev. Biochem., 56:365-394.

THERAPEUTIC STRATEGIES EMPLOYING CD4, THE HIV RECEPTOR

Per Ashorn, Bernard Moss, and Edward A. Berger

Laboratory of Viral Diseases
National Institute of Allergy and Infectious Diseases
National Institutes of Health
Bethesda, MD 20892

INTRODUCTION

Therapeutic Concepts Based on CD4

Infection of T-lymphocytes and macrophages by human immunodeficiency virus (HIV) is initiated by binding of the external subunit of the viral envelope glycoprotein (gp120) to CD4 molecules on the target cell membrane (1). Recombinant soluble forms of CD4 retain the capacity for high affinity binding to gp120 (2-7), suggesting potential therapeutic uses of CD4 derivatives. The diverse concepts for exploiting the CD4/gp120 interaction for therapy are outlined in Figure 1. The initial concept was based on the finding that soluble truncated forms of CD4 (sCD4) are able to neutralize HIV infectivity *in vitro* (2-6) (Figure 1A). Several "second generation" applications of this neutralizing concept are under development, including attachment of the CD4 to immunoglobulin constant region sequences (8-12), or presentation of the CD4 in association with erythrocytes (13). These modifications provide the advantages of increased plasma half-life compared to sCD4, and possibly enhanced efficiency of neutralization due to multivalency. Recent findings indicate that sCD4 is capable of "stripping" gp120 from the envelope glycoprotein complex (14-16), suggesting that the mechanism of sCD4 neutralization of HIV infectivity might involve more than simple competition for gp120 binding to CD4 on the surface of the target cell.

A second concept is to construct hybrid molecules containing the gp120-binding region of CD4 linked to a cytotoxic effector molecule (Figure 1B). In this strategy, the CD4 moiety serves as a targeting agent to direct the hybrid molecule to selectively kill HIV-infected cells which express the envelope glycoprotein at their surface (Figure 2). This has been achieved with a recombinant protein containing a portion of CD4 linked to active regions of *Pseudomonas* exotoxin A (17). Alternatively, sCD4 has been chemically conjugated to ricin A chain (18), and a recombinant molecule containing CD4 sequences linked to diphtheria toxin fragment A has been expressed (19). The aforementioned CD4-immunoglobulin hybrid proteins also have the potential to selectively kill HIV-infected cells by virtue of the Fc regions which can mobilize normal immune system functions, such as complement-mediated lysis and antibody dependent cellular cytotoxicity (ADCC).

Innovations in Antiviral Development and the Detection of Virus Infection
Edited by T. Block *et al.*, Plenum Press, New York, 1992

A third concept involves delivery systems which rely on binding between gp120 and membrane-associated CD4 (Figure 1C). Liposomes containing recombinant CD4 have been prepared (20), thus offering a means to selectively deliver cytotoxic compounds to HIV-infected cells. A most intriguing strategy is the development of gp120-based retroviral vectors (21) to target anti-HIV genetic sequences to CD4-positive cells (e.g. antisense sequences (22) or ribozymes (23)). Such sequences potentially could either prevent establishment of infection after virus entry into a target cell, or inhibit virus expression in cells already containing an integrated proviral genome. The high affinity interaction between CD4 and gp120 has thus spawned a diversity of strategic concepts for treatment of HIV infection. This report focuses on one of these approaches.

A. Direct Neutralization of HIV Infection by CD4 Derivatives

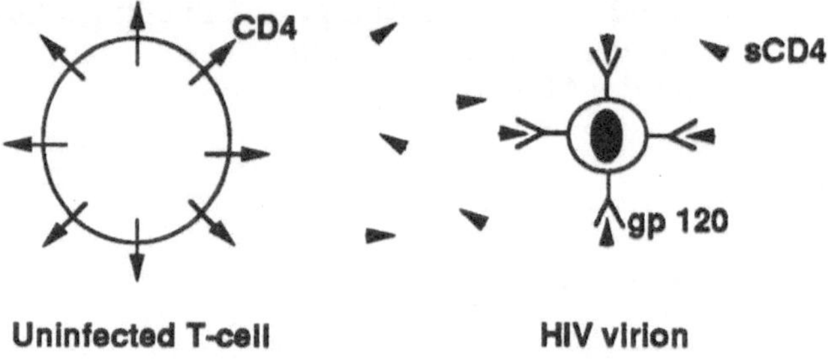

B. Selective Killing of HIV-Infected Cells by CD4-Hybrid Proteins

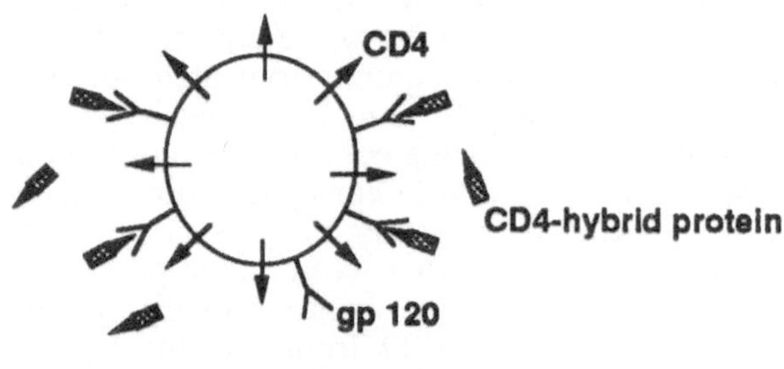

C. CD4/gp120-Based Delivery Systems

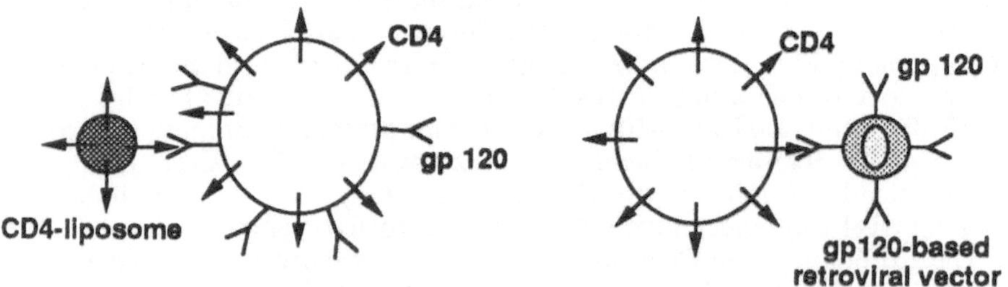

Figure 1. Therapy concepts based on the CD4/gp120 interaction.

CD4-*Pseudomonas* Exotoxin Hybrid Protein

Pseudomonas aeruginosa exotoxin A (PE) is a 66 kD monomeric protein in which specific structural domains are associated with specific functions (Figure 3). This domain structure allows the design of genetically engineered derivatives containing alternative ligands in place of the normal cell binding domain (24). We expressed in *E. coli* recombinant hybrid toxins consisting of portions of CD4 linked to the translocation and ADP-ribosylation domains of PE. The molecule we have focused on, designated CD4 (178)-PE40, contains the first two domains of CD4, including the gp120-binding site located within the amino-terminal domain. CD4(178)-PE40 selectively binds to cells expressing gp120 (17), and is presumably translocated to the cytoplasm where it ADP ribosylates elongation factor 2, resulting in inactivation of protein synthesis and consequent cell death.

We have used two types of systems to assess the anti-HIV activity of CD4(178)-PE40 (17, 25-28). In the first, we measure the ability of the toxin to kill a cell population in which most of the cells express surface HIV envelope glycoprotein. This can be achieved using cells expressing recombinant envelope glycoproteins, e.g. by using vaccinia virus vectors (17, 25); alternatively, cell lines which are chronically infected with HIV can be tested as targets (17, 25-28). In the second system, we add CD4(178)-PE40 to a population of susceptible target cells containing only a small number of infected cells, and test its ability to inhibit spread of HIV throughout the population (25, 27, 28). We can monitor the hybrid toxin's ability to protect a susceptible population from virus-mediated killing, as well as to suppress virus production. It is essential to include controls to distinguish whether the protective effects observed are truly due to selective killing of the infected cells, and not merely to a simple neutralization effect by the CD4 moiety of the hybrid toxin. In both the direct killing and HIV spread systems, MTT oxidation has served as a convenient, highly reproducible measure of the relative numbers of viable cells.

RESULTS AND DISCUSSION

Selective Killing of HIV-Infected Cells by CD4(178)-PE40

We have examined the cytotoxic activity of CD4(178)-PE40 against two chronically infected T-cell lines that constitutively produce virus (17, 25, 26). Table 1 shows that the hybrid toxin kills both cell lines with IC_{50} values in the range of 100 pM, but has no effect on the corresponding uninfected parental cell lines. The resistance of the parental lines to CD4(178)-PE40 is not simply due to their inherent insensitivity to PE derivatives, since native PE efficiently kills both the infected and uninfected cells (not shown). Control experiments verify that the specificity of CD4(178)-PE40 for HIV-infected cells is provided by the CD4 moiety: 1. PE40, which lacks a cell binding domain, has no effect on infected or uninfected cells (Table 1), and 2. the activity of the hybrid toxin is neutralized by anti-CD4 monoclonal antibodies, by recombinant gp120, and by sCD4, none of which affect native PE (26). CD4 derivatives lacking an enzymatically active domain [e.g. sCD4 or CD4(178)-PE40asp553] fail to promote killing (Table 1), demonstrating that the selective killing of HIV-infected cells by CD4(178)-PE40 is not due simply to occupation of the CD4 binding sites on gp120.

It is believed that individuals infected with HIV harbor reservoirs of latently infected cells, and that activation of virus production in such cells may play a critical role in the transition from the asymptomatic to the diseased state (29). We therefore tested the CD4(178)-PE40 sensitivity of latently HIV-infected cell lines of both the T-lymphocyte (26) and monocyte/macrophage (28) lineages. For each cell line in the

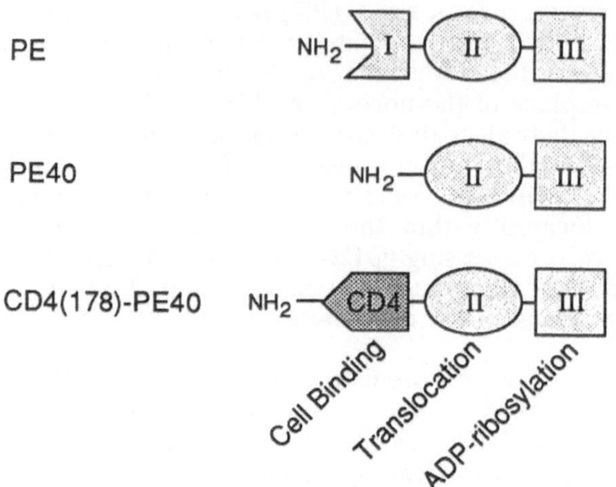

Figure 2. Selective killing of HIV-infected cells by a hybrid protein containing CD4 linked to a toxin. The CD4 moiety directs the hybrid toxin to selectively bind to and kill the HIV-infected cell, which expresses gp120 at its surface. The uninfected cell lacks gp120 and is thus spared from killing.

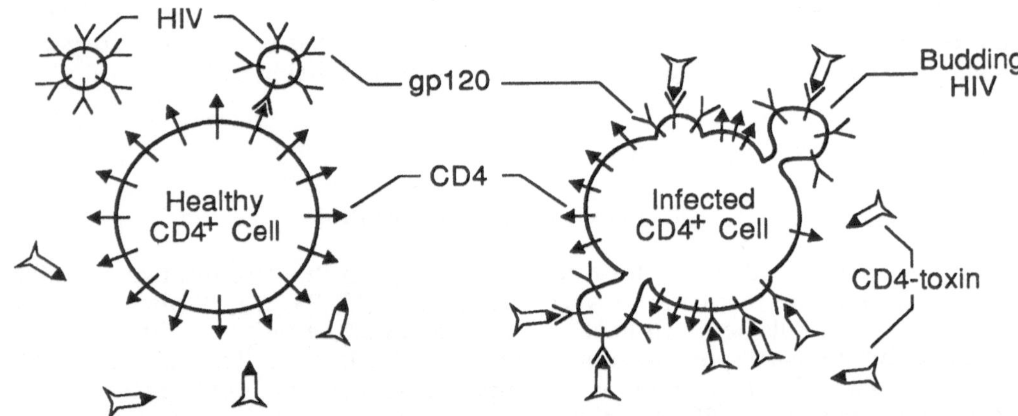

Figure 3. Structure/function domains of PE and its deriviatives. Domain I of native PE binds to receptors present on a wide variety of mammalian cell types, giving the toxin broad specificity; domain II is involved in translocation of the toxin to the cytoplasm; domain III ADP-ribosylates elongations factor 2, thereby inactivating protein synthesis and killing the cell. PE40 is a recombinant construct lacking domain I, and serves as a control for non-specific toxicity. In CD4(178)-PE40, the CD4 region consists of the N-terminal 178 amino acids, representing the first two domains of CD4; the gp120-binding site is contained within the first domain.

basal state, only a small fraction of cells express low levels of virus; however, induction leads to virus expression in most of the cells. Figure 4 shows that for both latently infected lines, the cells are readily killed by the hybrid toxin, but only when virus expression is induced; by contrast, the parental uninfected cell lines are not killed under either condition. These results have obvious clinical implications for the therapeutic use of CD4(178)-PE40, since they suggest that the hybrid toxin will only eliminate the sub-population of infected cells actively producing virus. This may influence the timing of treatment protocols involving this agent.

Inhibition of Spreading HIV Infection by CD4(178)-PE40

The therapeutic potential of CD4(178)-PE40 depends on its ability to inhibit HIV spread throughout a susceptible cell population by selectively killing the infected cells. We have developed several systems to study this problem, including continuous T-cell lines (25, 27) and primary cultures of human T-cells (27, 28) and macrophages (28). Table 2 summarizes the results indicating that the hybrid toxin greatly suppresses virus production in each of these systems. Furthermore, in both the continuous (25, 27) and primary (28) T-cell cultures, the virus-mediated killing of the target cell population is significantly delayed (not shown). However, we observe that the infection does eventually spread and eliminates most of the target cell population (25, 27). These results are entirely consistent with the mode of action of the toxin. CD4(178)-PE40 can only kill an infected cell once sufficient envelope glycoprotein has accumulated at the cell surface. Complete inhibition of target cell killing would be expected only if the newly infected cell is killed by the toxin before any virus spread occurs. Apparently, some virus spread does occur before the infected cell is killed by the toxin; as a result the infection eventually spreads to all target cells. Because the infected cells are killed by the toxin before producing the quantity of virus they would normally produce before succumbing to the viral cytopathic effect, the total yield of virus is greatly suppressed, and cell killing is delayed. Control proteins, including soluble CD4 or CD4(178)-PE40asp533, are much less effective than CD4(178)-PE40 in this system (Table 2, Refs 25, 27, 28), indicating that most of the protective effect observed with the hybrid toxin is due to selective killing of the infected cells, and not merely to a neutralization effect by the CD4 moiety of the toxin.

Synergistic Action of CD4(178)-PE40 and Reverse Transcriptase (RT) Inhibitors

It is becoming increasingly appreciated that effective treatment of HIV infection will probably require combinations of drugs that act at different aspects of the infection process(30). In considering this concept, we felt that a particularly potent combination would involve the use of a drug which blocks viral replication and resulting spread of infection to new target cells, plus another which kills those cells already infected. Examples of the former type (virostatic drugs) are RT inhibitors such as AZT and ddI (31); an example of the latter type (virocidal drug) is CD4(178)-PE40. Strong synergistic effects between CD4(178)-PE40 and RT inhibitors are noted in PBMC cultures acutely infected with HIV (27). Each class of drug strongly potentiates the activity of the other; that is, in the presence of one drug, the IC_{50} of the other drug is reduced several fold. These findings prompted us to test the ability of combinations of CD4(178)-PE40 and RT inhibitors to eliminate infectious HIV from cell cultures (27). Figure 5 shows an experiment with cultures of a human T-cell line acutely infected with HIV. The experiment is divided into two phases: a 31-day treatment phase during which the indicated drugs are continually replenished, and a subsequent phase after termination of drug treatment. Protection against HIV

Table 1. Sensitivity of HIV-1 infected and uninfected T-cell lines to killing by CD4(178)-PE40 or control proteins. Two T-cell lines chronically infected with HIV-1 were tested; the corresponding uninfected parental cell lines served as controls. After 4-5 days of culture in the presence of the indicated proteins, the relative numbers of viable cells were determined by the MTT assay. IC_{50} is the concentration giving 50% cell killing. CD4-PE40asp533 is a recombinant protein with greatly reduced cytotoxicity due to a mutation abolishing ADP-ribosylation activity. In some experiments lys-PE40, a genetically engineered version of PE40 with an additional lysine residue at the N-terminus, was used instead of PE40. * = no toxicity at this concentration; n.d. = not determined.

		IC_{50} (nM)			
Cell Line	HIV-1 Infected	CD4-PE40	PE40	sCD4	$CD4\text{-}PE40_{asp533}$
8E5	YES	0.08	>10*	>10*	>10*
A3.01	NO	>10*	>10*	>10*	>10*
H9/HTLVIIIB	YES	0.1	>8*	n.d.	n.d.
H9	NO	>10*	>8*	n.d.	n.d.

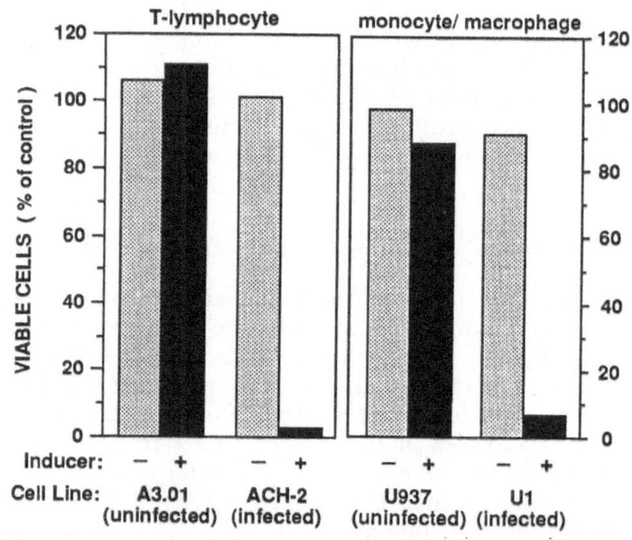

Figure 4. Effects of CD4(178)-PE40 on chronically infected cell lines inducible for HIV-1 expression. The T-lymphocyte lines were cultured 5 days with 4 nM CD4(178)-PE40, in the absence or presence of inducer (a crude cytokine mix with TNF-alpha as the major active component). The monocyte/macrophage lines were treated for 24 hours without or with inducer (10 nM PMA), then cultured 4 days with 100 nM CD4(178)-PE40. Relative viable cell numbers are expressed as the % of the MTT values obtained for the corresponding cultures in the absence of CD4(178)-PE40.

Table 2. Inhibition of virus production by CD4(178)-PE40 during spreading HIV infection in various cell types. The effects of the indicated proteins are expressed as the % of total virus produced in control cultures with no additions. For the T-cell line experiment (25), chronically infected H9/HTLV IIIB cells were mixed with uninfected A3.01 cells at a ratio of 1:1000. Cultures were maintained in the absence or presence of the indicated CD4 derivative at 10 nM. Every two days the RT activity was measured and the cultures were split 1:5. The results shown are based on the total RT measured during 18 days, at which point nearly all of the target cell population had been eliminated. For the primary T-lymphocyte experiment (28), PHA-stimulated PBMCs were infected with HIV-1 (LAV isolate) and cultured with IL-2 in the absence or presence of the indicated CD4 derivative at 50 nM. The results shown are based on the p24 produced from days 3-6, which represented at least 80% of the total which could be produced in this system. For the primary macrophage experiment (28), primary monocyte/macrophages obtained by elutriation were stimulated with recombinant CSF-1. The adherent cells were infected with HIV-$1_{AD-87\ (M)}$, then cultured in the absence or presence of the indicated CD4 derivate at 10 nM. The results shown are based on the total RT produced during 15 days.

	Virus produced (% of control)	
Human Target Cell	CD4(178)-PE40	sCD4
T-cell line (A3.01)	7	55
Primary T-lymphocytes	8	95
Primary macrophages	17	120

spread is monitored by inhibition of virus-mediated killing of the target cell population and suppression of virus production. In the absence of drugs, the target cell population is killed off by the second week of culture, with a concomitant peak of virus production. As noted above, CD4(178)-PE40 alone delays, but does not prevent, cell killing; some inhibition of virus production is observed. AZT completely blocks virus spread as measured by either parameter, but only while present in the culture. Once AZT treatment is terminated, cell killing and virus production rapidly ensue, presumably because of the persistence of infected cells in the population. The inclusion of soluble CD4 along with AZT does not enhance the antiviral effect. However, combination of CD4(178)-PE40 and AZT completely inhibits virus spread, both during and after the drug treatment phase. Furthermore, quantitative PCR analysis, sufficiently sensitive to detect provirus from a single infected cell, indicates the complete absence of HIV DNA after the combined treatment protocol. Control experiments (not shown) indicate that the treatment does not merely select for CD4-negative, HIV-resistant cells, since the surviving cell population is CD4-positive by immunofluorescence, and is efficiently killed by the addition of fresh HIV. We conclude that the combined treatment with CD4(178)-PE40 plus AZT completely eliminates infectious HIV from this culture system.

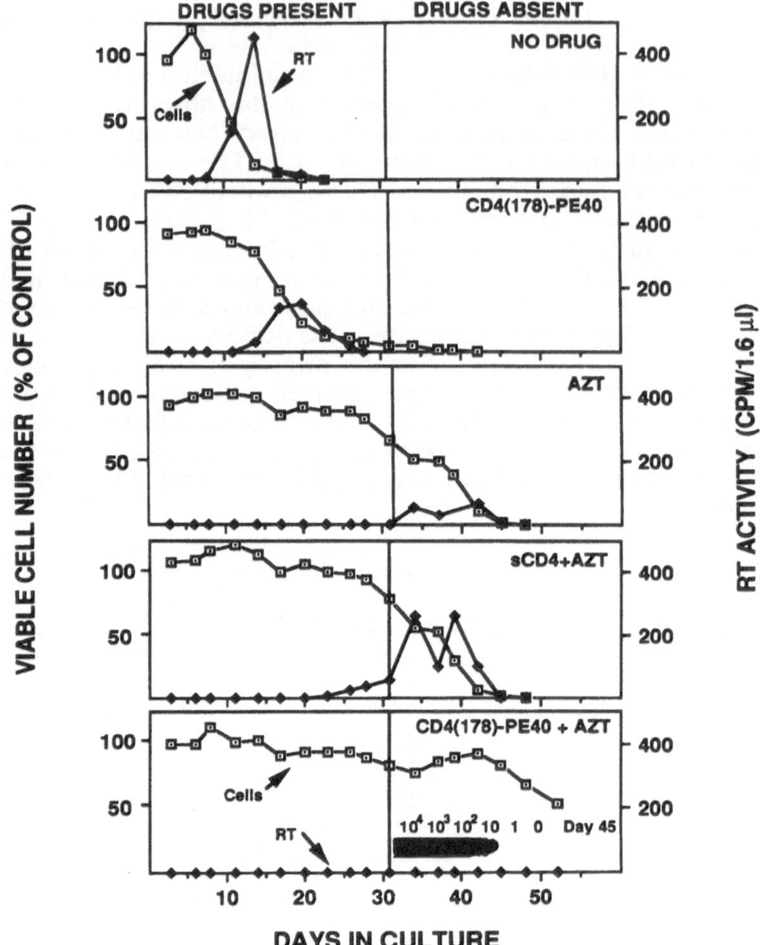

Figure 5. Elimination of infectious HIV by combination treatment with CD4(178PE40 at plus AZT. A3.01 cells were acutely infected with HIV-1 (LAV isolate), then cultured in the presence of the indicated drugs (CD4(178)-PE40 and sCD4 at 10 nM; AZT at 1 μM). The cultures were diluted 1:12 every 2-3 days with medium containing drugs at the original concentrations. Beginning at day 31, the drugs were omitted from the dilution medium. Virus production was monitored by RT activity in the supernatant. Relative viable cell numbers were determined by the MTT procedure, and are expressed as the percent of the value obtained with a control culture containing no HIV. The gradual decline in relative cell number after cessation of drug treatment in the lowest panel is due to variation introduced by the repeated dilution process, not to HIV spread. The autoradiograph in the lowest panel shows the assay for HIV-1 DNA using a PCR-amplification method. On the left is a series of signals from a standard series of reaction mixtures, each containing the indicated numbers of HIV-1 proviruses (obtained by serial dilution of a lysate from chronically infected U1 cells). The experimental sample on the right shows the absence of a signal using a volume of lysate corresponding to 75% of the cells remaining at day 45. See Reference 27 for details.

These results highlight the therapeutic potential of combination drug therapy involving a virostatic agent capable of blocking HIV replication and consequent infection of new target cells, plus a virocidal compound which kills those cells already infected.

CONCLUSIONS AND FUTURE DIRECTIONS

The finding presented herein demonstrate the potent anti-HIV properties of CD4(178)-PE40 *in vitro*. Several additional factors must be considered in extrapolating these results to the clinical situation. First, it must be noted that all our studies to date have been conducted with common laboratory HIV isolates which have been propagated for long periods *in vitro*. In view of the recent report that primary isolates from infected individuals are much less sensitive than laboratory isolates to neutralization by sCD4 (32), it is of great importance to test the effects of CD4(178)-PE40 against fresh patient isolates. Furthermore, the report that sera from some infected individuals contains anti-gp120 antibodies capable of inhibiting binding of sCD4 (33) makes it imperative to test for possible neutralizing activity against CD4(178)-PE40 in patient sera. Problems arising from immunogenicity and non-specific toxicity, as previously observed in animals with another PE derivative (34), may restrict the time period over which the drug can be administered. Nevertheless, the promising results obtained thus far with CD4(178)-PE40 suggest that the hybrid toxin, particularly in combination with other anti-HIV agents, could potentially eliminate a large fraction of infected cells in the body, thereby reducing the viral burden and providing significant therapeutic effect to HIV-infected individuals.

ACKNOWLEDGEMENTS

We thank P. B. Robbins for excellent technical assistance. This work was supported in part by the National Institutes of Health Intramural AIDS Targeted Antiviral Program.

REFERENCES

1. Klatzman, D.R., McDougal, J.S., Maddon, P.J., 1990, the CD4 Molecule and HIV infection, Immunodef. Rev. 2:43
2. Smith, D.H., Byrn, R.A., Marsters, S.A., Gregory, T., Groopman, J.E., Capon, D.J., 1987, Blocking of HIV-1 infectivity by a soluble, secreted form of the CD4 antigen, Science 238:1704.
3. Fisher, R.A., Bertonis, J.M., Meier, W., Johnson, V.A., Costopoulos, D.S., Liu, T., Tizard, R., Walker, B.D., Hirsch, M.S., Schooley, R.T., Flavell, R.A., 1988, HIV infection is blocked in vitro by recombinant soluble CD4, Nature 331:76.
4. Hussey, R.E., Richardson, N.E., Lowalski, M., Brown, N.R., Chang, H.-C., Siliciano, R.F., Dorfman, T., Walker, B., Sodroski, J., Reinherz, E.L., 1988, a soluble CD4 protein selectively inhibits HIV replication and syncytium formation, Nature 331:78.
5. Deen, K.C., McDougal, J.S., Inacker, R., Folena-Wasserman, G., Arthos, J., Rosenberg, J., Maddon, P.J., Axel, R., Sweet, R.W., 1988, a soluble form of CD4 (T4) protein inhibits AIDS virus infection, Nature 331:82.
6. Traunecker, A., Luke, W., Karjalainen, K., 1988, Soluble CD4 molecules neutralize human immunodeficiency type 1, Nature 331:84.

7. Berger, E.A., Fuerst, T.R., Moss, B., 1988, a soluble recombinant polypeptide comprising the amino-terminal half of the extracellular region of the CD4 molecule contains an active binding site for human immunodeficiency virus, Proc. Natl. Acad. Sci., USA 85:2357.
8. Capon, D.J., Chamow, S.M., Mordenti, J., Marsters, S.A., Gregory, T., Mitsuya, H., Byrn, R.A., Lucas, C., Wurm, F.M., Groopman, J.E., Smith, D.H., 1989, Designing CD4 immunoadhesions for AIDS therapy, Nature 337:525.
9. Traunecker, A., Schneider, J., Kiefer, H., Karjalainen, K., 1989, Highly efficient neutralization of HIV with recombinant CD4-immunoglobulin molecules, Nature 339:68.
10. Mizukami, T., Smith, C.D., Berger, E.A., Moss, B., 1989, Expression and characterization of chioneric proteins containing human CD4 linked to human immunoglobulin heavy chain constant regions, V. Int. Conf. AIDS Abst. M.C. p.89, p.556.
11. Byrn, R.A., Mordenti, J., Lucas, C., Smith, D., Marsters, S.A., Johnson, J.S., Cossum, P., Chamow, S.M., Wurm, F.M., Gregory, T., Groopman, J.E., Capon, D.J., 1990, Biological properties of a CD4 immunoadhesius, Nature 344:667.
12. Zettlmeissl, G., Gregersen, J.P., Duport, J.M., Mehdi, S., Reiner, G., Seed, B., 1990, Expression and characterization of human CD4 immunoglobulin fusion proteins, DNA Cell. Biol. 9:347.
13. Mouneimne, Y., Tosi, P.F., Barhoumi, R., Nicolau, C., 1990, Electroinsertion of full length recombinant CD4 into red blood cell membrane, Biochim. Biophys. Acta 1027:53.
14. Kirsh, R., Hart, T.K., Ellens, H., Miller, J., Petteway, S.A., Lambert, D.M., Leary, J., Bugelski, P.J., 1990, Morphometric analysis or recombinant soluble CD4 mediated release of the envelope glycoprotein gp120 from HIV-1, AIDS Res. Hum. Retroviruses 6:1209.
15. Moore, J.P., McKeating, J.A., Weiss, R.A., Sattentau, Q.J., 1990, Dissociation of gp120 from HIV-1 virions induced by soluble CD4, Science 250:1139.
16. Berger, E.A., Lifson, J.D., Eiden, L.E., 1991, Stimulation of glycoprotein gp120 dissociation from the envelope glycoprotein complex of human immunodeficiency virus type 1 by soluble CD4 and CD4 peptide derivatives: implications for the role of the complementarity - dtermining region 3-like region in membrane fusion, Proc. Natl. Acad. Sci USA 88:8082.
17. Chaudhary, V.K., Mizukami, T., Fuerst, T.R., FitzGerald, D.J., Moss, B., Pastan, I., Berger, E.A., 1988, Selective killing of HIV-infected cells by recombinant human CD4-Pseudomonas exotoxin hybrid protein, Nature 333:369.
18. Till, M.A., Ghetie, V., Gregory, T., Patzer, E.J., Porter, J.P., Uhr, J.W., Capon, D.J., Vitetta, E.S., 1988, HIV-infected cells are killed by rCD4-ricin A chain, Science 242:1166.
19. Zverev, V., Malushova, V., Sidorov, A., Zdanovsky, A., Blinov, V., Korneeva, M., Pugach, A., Yankovsky, N., Andjaparidze, O., 1990, Inhibition of the infectivity of HIV by different recombinant forms of CD4, VI Int. Conf. AIDS Abst. th. A. 252, vol.1, p. 183.
20. Cudd, A., Noonan, C.A., Tosi, P.F., Melnick, J.L., Nicolau, C., 1990, Specific interaction of CD4-bearing liposomes with HIV-infected cells, J. Acquir. Immune Defic. Syndr. 3:109.
21. Lever, A., Richardson, J., Gottlinger, H., Haseltine, W.A., Sodroski, J., 1990, Development of a retroviral vector system based on HIV, VI Int. Conf. AIDS Abst. S.A. 349, vol. 3, p. 169.
22. Agrawal, S., Ikeuchi T., Sun, D., Sarin, S., Konopka, A., Maizel, J., Zamecnik, P.C., 1990, Inhibition of human immunodeficiency virus in early infected and chronically infected cells by antisense oligodeoxynucleotides and their phosphorothioate analogues, Proc. Natl. Acad. Sci., USA 86:7790.

23. Sarver, N., Cantin, E.M., Chang, P.S., Zaia, J.A., Ladne, P.A., Stephens, D.A., Rossi, J.J., 1990, Ribozymes as potential anti-HIV-1 therapeutic agents, Science 247:1222.
24. Pastan, I., FitzGerald, D., 1989, Pseudomonas exotoxin: Chimeric Toxins, J. Biol. Chem. 264:15157.
25. Berger, E.A., Clouse, K.A., Chaudhary, V.K., Chakrabarti, S., FitzGerald, D.J., Pastan, I., Moss, B., 1989, CD4-Pseudomonas exotoxin hybrid protein blocks the spread of human immunodeficiency virus infection *in vitro* and is active against cells expressing the envelope glycoproteins from diverse primate immunodeficiency retroviruses, Proc. Natl. Acad. Sci. USA 86:9539.
26. Berger, E.A., Chaudhary, V.K., Clouse, K.A., Jaraquemada, D., Nicholas, J.A., Rubino, K.L., FitzGerald, D.J., Pastan, I., Moss, B., 1990, Recombinant CD4-Pseudomonas exotoxin hybrid protein displays HIV-specific cytotoxicity without affecting MHC-class II-dependent functions, AIDS Res. Hum. Retroviruses 6:795.
27. Ashorn, P., Moss, B., Weinstein, J.N., Chaudhary, V.K., FitzGerald, D.J., Pastan, I., Berger, E.A., 1990, Elimination of infectious human immunodeficiency virus from human T-cell cultures by synergistic action of CD4-Pseudomonas exotoxin and reverse transcriptase inhibitors, Proc. Natl. Acad. Sci. USA 87:8889.
28. Ashorn, P, Englund, G., Martin, M.A., Moss, B., Berger, E.A., 1991, Anti-HIV activity of CD4-Pseudomonas exotoxin on infected primary human lymphocytes and monocyte/macrophages, J. Infect. Dis., 163:703.
29. Rosenberg, Z.F., Fauci, A.S., 1990, Immunopathogenic mechanisms of HIV infection: Cytokine induction of HIV expression, Immunology Today 11:176.
30. Johnson, V.A., Hirsh, M.S., 1989, Antiviral Chemotherapy: New directions for clinical applications and research, J. Mills and L. Corey, eds., Elsevier, New York, 2:275.
31. Yarchoan, R., Mitsuya, H., Myers, C.E., Broder, S., 1989, Clinical pharmacology of 3'-azido 2', 3'-dideoxythymidine (zidovudine) and related dideoxynucleosides, N. Engl. J. Med. 321:726.
32. Daar, E.S., Li, X.L., Moudgil, T., Ho, D.D., 1990, High concentrations of recombinant soluble CD4 are required to neutralize primary human immunodeficiency virus type 1 isolates, Proc. Natl. Acad. Sci., USA 87:6574.
33. Callahan, L.N., Norcross, M.A., 1989, Inhibition of soluble CD4 therapy by antibodies to HIV (letter), Lancet 734.
34. Loberboum-Galski, H., Barrett, L.V., Kirkman, R.L., Ogata, M., Willingham, M.C., FitzGerald, D.J., Pastan, I., 1989, Cardiac allograft survival in mice treated with IL2-PE40, Proc. Natl. Acad. Sci. USA 86:1008.

MOLECULAR CHARACTERIZATION OF HIV-2 (ROD) PROTEASE FOLLOWING PCR CLONING FROM VIRUS INFECTED H9 CELLS

Y.-S. Edmond Cheng, Catherine E. Patterson, Ronald G. Rucker
Michael J. Otto, Christopher J. Rizzo, and Bruce D. Korant

DuPont Merck Pharmaceutical Company
Wilmington, DE 19880-0328

SUMMARY

A 450 nucleotide sequence corresponding to the nucleotides 1931-2380 of the viral genome (8) was amplified by polymerase chain reaction (PCR) using template DNA prepared from HIV-2 (ROD) infected H9 cells. The sequence codes for HIV-2 protease and its N-terminal flanking peptide. An identical DNA sequence was obtained from three independent PCR amplifications, which differs from the published sequence of HIV-2 (ROD) in 7 nucleotides scattered throughout the region of the cloned DNA. The cloned DNA was expressed in *E. coli* cells and resulted in the synthesis of a correctly processed HIV-2 protease, which is enzymatically active. Therefore, none of the seven nucleotide changes, which resulted in two amino acid substitutions, affect the autoproteolytic or trans-cleaving activities of the HIV-2 protease.

INTRODUCTION

Human immunodeficiency virus (HIV) type 1 and 2 are etiological agents for the acquired immunodeficiency syndrome (AIDS). HIV codes for an aspartyl protease essential for the proteolytic processing of the *gag* and *gag-pol* polyproteins (1,2). This protease has therefore been recognized as a target enzyme for the development of antivirals. The crystal structure of HIV-1 protease (HIV-1 PR) and its inhibitor complexes have been solved (3,4). Molecular modeling and enzymatic comparisons suggested that the HIV-2 protease (HIV-2 PR) is structurally and functionally similar, but nonidentical to HIV-1 PR (5,6,7). Structural analysis of crystals of HIV-2 PR and its inhibitor complexes would be help for detailed comparisons with HIV-1 PR and in the design of antivirals for HIV 1 and 2.

This report describes a rapid method to obtain the HIV-2 protease sequence from virus infected H9 cells, and to express it in bacterial cells.

MATERIALS AND METHODS

Materials

HIV-2, ROD isolate (8), was obtained from David Montefiori (Vanderbilt University). *E. coli* K12 cells, JM105, the expression vector, pET3AM, and the rabbit antibody for HIV-1 PR were described previously (9,10).

PCR Amplification of a 450 nt DNA Containing the Sequence for the HIV-2 PR Precursor

Human H9 cells were infected with HIV-2 (ROD isolate) and harvested either 3 days or 3 months (chronically infected) after infection. Total DNA preparations were made by phenol extraction of the harvested cells and used as a template for the polymerase chain reactions. Two oligonucleotides, TCGAATTCATGTCCAGCAGTGGATCTACTGG, and TGAAGCTTACTATAGATTTAATGACATGCCTAAG were used as PCR primers to amplify the 450 nt DNA, which corresponds to nucleotides 1931-2380 of the HIV-2 (ROD) viral genome sequence described by Guyader et al.(8). The amplified DNA was cloned into pTZ18R (Pharmacia) and sequenced.

Expression of the HIV-2 PR in *E. coli* cells

The DNA insert was recloned into the expression vector, pET3AM, at the *Eco*RI and *Hin*DIII sites, introduced into JM105 cells and expressed as described (9).

Purification of the HIV-2 PR

E. coli cells induced for the production of HIV-2 PR were harvested and lysed by passage through a French press cell. HIV-2 PR in cell lysates were clarified by centrifugation at 45,000g and purified using pepstatin agarose and Mono-S columns as described (11).

RESULTS AND DISCUSSION

Sequence of HIV-2 PR clones isolated from virus infected H9 cells

We have prepared total DNA samples from either 3-day or 3-month HIV-2 infected H9 cultures. These DNA samples were amplified in three separate PCR reactions, and the resulting amplified DNA cloned into pTZ18R plasmid. Three clones, each from an independent reaction, were isolated and completely sequenced. An identical sequence was obtained from the three independently isolated clones. This sequence consists of 450 nucleotides spanning 1931-2380 of the HIV-2 (ROD) *pol* gene. The sequence we obtained contains 7 single nucleotide changes from the published sequence (Figure 1). Two of the changes, at nt's 2009 (A-G) and 2040 (G-A), result in amino acid changes from Ser to Gly and from Gly to Glu respectively. The other 5 single nucleotide changes: nt's 2149 (C-T), 2161 (A-G), 2164 (G-A), 2200 (G-A), and 2272 (A-G) result in no amino acid changes.

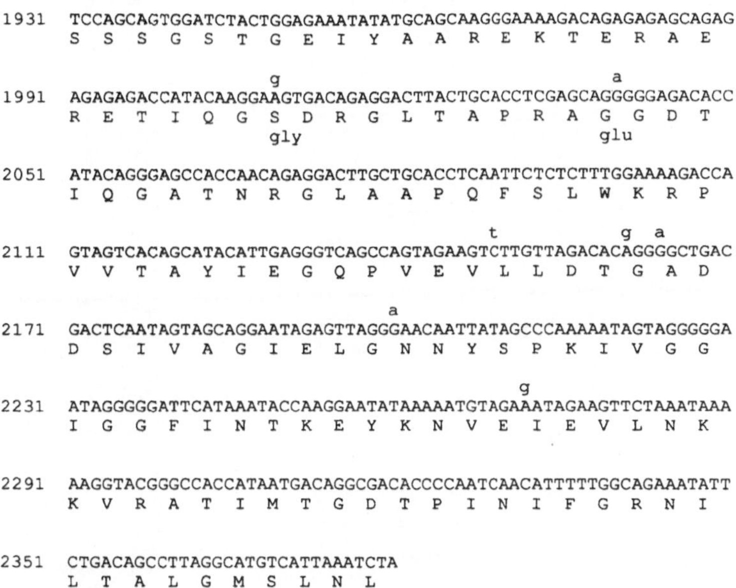

Figure 1. Mutations at the 5'-end of the *pol* gene of HIV-2 (ROD) virus. The DNA sequence and the deduced amino acid sequence of part of the *pol* gene (nt 1931-2380) are shown according to Guyader et al., (8). Single-nucleotide mutations present in this region in HIV-2 infected H9 cells are shown on the top line, and the corresponding amino acid changes are shown on the bottom line. The nucleotide numbers are shown on the left column.

Expression of HIV-2 Protease from the PCR Amplified DNA

The two altered amino acids are located "upstream" to the autoproteolytic cleavage site. To determine the effect of these two altered amino acids on the autoproteolytic process, we cloned the entire 450 nt insert to examine if the expressed precursor protein can be properly processed to the mature 99-AA protease. The recombinant protease made in *E. coli* is soluble. It is also correctly cleaved because the expected N-terminal sequence, Pro-Gln-Phe-Ser-Leu-Trp-Lys-Arg-Pro-Val. . ., of the mature 99-AA protease (5,6) was detected at the amino terminus of the isolated protein. Therefore, the two amino acid changes we found in the N-terminal flanking peptide of HIV-2 have no apparent effect on HIV-2 PR processing.

Activity of Purified HIV-2 PR

After an affinity purification using pepstatin agarose, the HIV-2 PR still contained several contaminating proteins that could not be removed by repeated pepstatin affinity chromatography (Figure 2, lane 3). Additional column chromatography using a Mono-S column completed the purification (Figure 2, lane 4).

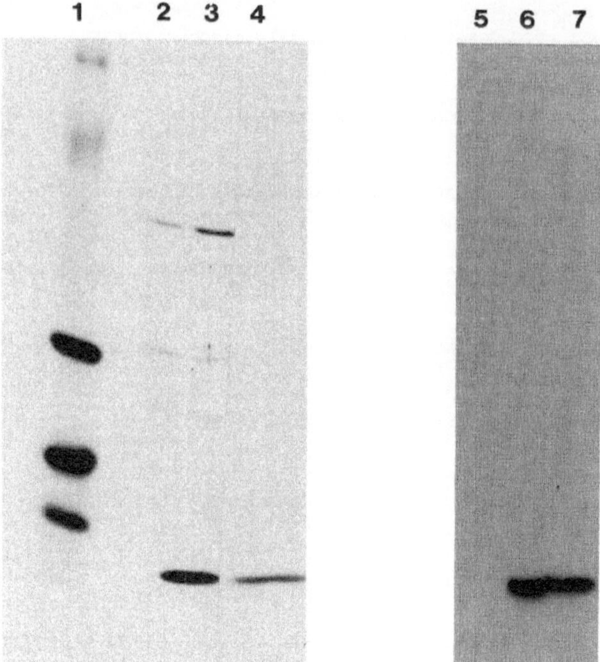

Figure 2. Column purified HIV-2 protease. Soluble cell extracts of a control JM105 culture (lane 2 and 5) or a protease expression culture (lanes 3,4,6,7) were purified with pepstatin agarose (lanes 2,3,6) or with both pepstatin agarose and Mono-S columns (lanes 4,7) and separated by SDS-PAGE. Proteins were detected by Coomassie staining (lanes 1,2,3,4), and protease antigen detected by immunoblot (lanes 5,6,7). Protein size standards (6,000, 41,000, 29,000, 18,000, and 14,000 daltons) are shown in lane 1.

The isolated HIV-2 PR was recognized by a rabbit antibody for HIV-1 PR in a Western blot analysis (Figure 2, lanes 6-7) and was fully active in the proteolysis of the HIV-1 *gag* protein substrate (Figure 3). In addition, the purified protease is also active in the proteolysis of a peptide substrate, Ala-Thr-Val-Tyr-Phe(NO_2)-Val-Arg-Lys-Ala, with a k_{cat}/K_m of 7,000 +/- 1,400, comparable to that of HIV-1 PR (10).

The seven nucleotide changes within the 450 nt sequence of the HIV-2 are consistent with the reported high mutation frequency of retroviruses (12). The fact that all of the five mutations within the PR coding sequence are silent is an indication that protease structure is relatively intolerant of amino acid alterations. In contrast, the sequence in the protease precursor, amino terminal to the coding sequence, appears to accommodate certain amino acid changes without adverse effect on autoproteolysis. The mutations reported here were detected in HIV-2 infected cultures in the absence of selective pressure by antivirals. It is conceivable that additional mutations would be found in the viral progenies propagated in the presence of specific protease inhibitors such as those reported by Roberts et al., (13).

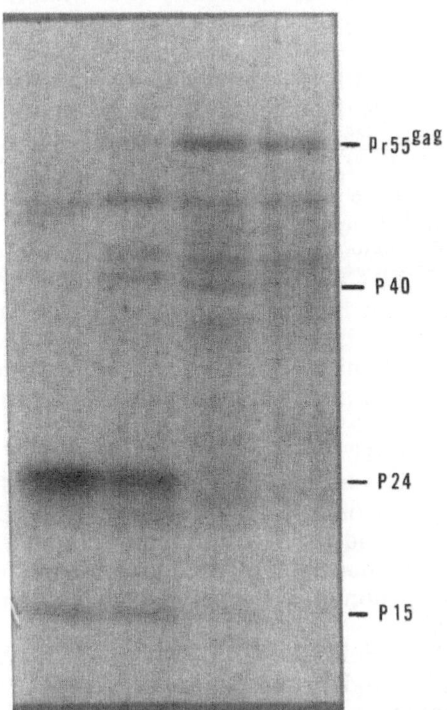

Figure 3. Proteolysis of HIV-1 *gag* by HIV-PR. Labelled HIV-1 *gag* protein substrate was incubated with 1:10 (1), 1:100 (2), 1:1000 (3) diluted HIV-2 PR as described (9), and the proteolysis of the $Pr55^{gag}$ was detected by autoradiography. Lane 4 is an undigested control. The position of $Pr55^{gag}$ and its processing products are shown at the right column.

ACKNOWLEDGEMENTS

We thank Dr. Charles Kettner for enzymatic assays and Dr. Ramnath Seetharam for N-terminal sequence determinations.

REFERENCES

Kramer, R.A., Schaber, M.D., Skalka, A.M., Ganguly, K., Wong-Staal, F., and Reddy E.P. 1986. HTLV-III gag protein is processed in yeast cells by the virus pol-protease. Science 231:1580-84.

Kohl, N.E., Emini, E.A., Schleif, W.A., Davis, L.J., Heimbach, J.C., Dixon, R.A.F., Scolnick, E.M., and Sigal, I.S. 1988. Active human immunodeficiency virus protease is required for viral infectivity. Proc. Natl. Acad. Sci. U.S.A. 85:4686-90.

Wlodawer, A., Miller, M., Jaskolski, M., Sathyanarayana, B.K., Baldwin, E., Weber, I.T., Selk, L.M., Clawson, L., Schneider, J., and Kent, S.B.H. 1989. Conserved folding in retroviral proteases: Crystal structure of a synthetic HIV-1 protease. Science 245:616-21.

Gustchina, A., and Weber, I.T. 1990. Comparison of inhibitor binding in HIV-1 protease and in non-viral aspartic proteases: The role of the flap. FEBS LETTERS 269:269-72.

Le Grice, S.F.J., Roswitha, E., Mills, J., and Mous, J. 1989. Comparison of the human immunodeficiency virus type 1 and 2 proteases by hybrid gene construction and trans-complementation. J. Biol. Chem. 364:14902-8.

Pichuantest, S., Babe, L.M., Barr, P.J., DeCamp, D.L., and Craik, C.S. 1990. Recombinant HIV2 protease processes HIV1 Pr53 gag and analogous junction peptides in vitro. J. Biol. Chem. 265:13890-8.

Tomasselli, A.G., Hui, J.O., Sawyer, T.K., Staples, D.J., Bannow, C., Reardon, I.M., Howe, W.J., DeCamp, D.L., Craik, C.S., and Heinrikson, R.L. 1990. Specificity and inhibition of proteases from human immunodeficiency viruses 1 and 2. J. Biol. Chem. 265:14675-83.

Guyader, M., Emerman, M., Sonigo, P., Clavel, F., Montagnier, L. and Alizon, M. 1987. Genome organization and transactivation of the human immunodeficiency virus type 2. Nature 326:662-9

Cheng, Y.-S.E., McGowan, M.H., Kettner, C.A., Schloss, J.V., Erickson-Viitanen S., and Yin, F.H. 1990. High level synthesis of recombinant HIV-1 protease and the recovery of active enzyme from inclusion bodies. Gene. 87:243-8.

Cheng, Y.-S.E., Yin, F.H., Founding, S., Blomstrom, D., and Kettner, C.A. 1990. Stability and activity of human immunodeficiency virus protease: Comparison of the natural dimer with a homologous, single-chain tethered dimer. Proc. Natl. Acad. Sci. U.S.A. 87:9660-4.

Rittenhouse, J., Turon, M.C., Helfrich, R.J., Albrecht, K.S., Weigl, D., Simmer, R.L., Mordini, F., Erickson, J., and Kohlbrenner, W.E. 1990. Affinity purification of HIV-1 and HIV-2 proteases from recombinant *E. Coli* strains using pepstatin-agarose. Biochem. Biophys. Acad. Comm. 171:60-6.

Domingo, E. 1989. RNA virus evolution and the control of viral disease. Prog. Drug Res. 33:93-133.

Roberts, N.A., Martin, J.A., Kinchington, D., Broadhurst, A.V., Craig, J.C., Duncan, I.B., Galpin, S.A., Handa, B.K., Kay, J., Krohn, A., Lambert, R.W., Merrett, J.H., Mills, J.S., Parkes, K.E.B., Redshaw, S., Ritchie, A.J., Taylor, D.L., Thomas, G.J., Machin, P.J. 1990. Rational design of peptide-based HIV proteinase inhibitors. Science 248:358-61.

A NOVEL, NON-NUCLEOSIDE INHIBITOR OF HIV-1 REVERSE TRANSCRIPTASE (REVIEW)

Vincent J. Merluzzi and Alan S. Rosenthal

Boehringer Ingelheim Pharmaceuticals, Inc.
Ridgefield, CT 06877

SUMMARY AND INTRODUCTION

Human immunodeficiency virus type 1 (HIV-1) is the retrovirus responsible for the majority (>95%) of the acquired immunodeficiency syndrome (AIDS) cases in the world. The complicated life-cycle of this virus presents many challenging areas for intervention. Our laboratories have concentrated on prevention of the early phase in proviral synthesis, specifically interruption of the RNA > DNA metabolic process by interfering with the viral enzyme, reverse transcriptase (RT). The RT of HIV-1 is a necessary component for early proviral synthesis. This enzyme has binding sites for nucleoside triphosphates, template primer and a catalytic site for the polymerase reaction. Nucleotides are added to the polymerizing chain to create a complementary DNA molecule (for review see Gilboa et al., 1979). The most effective inhibitors of RT have been the nucleoside analogs which are converted to triphosphates by cellular enzymes and act as chain terminators of the RT reaction (Mitsuya et al., 1985). Zidovudine (3'-azido-2'-3'dideoxy-thymidine, AZT) has been shown to be of benefit in HIV-1 infected individuals (Yarchoan et al., 1986). However, there are side-effects associated with the use of AZT (Richman et al., 1987), in addition to incomplete viral inhibition (Ho et al., 1989) and viral resistance (Larder et al., 1989).

From a series of compounds originally synthesized as muscarinic antagonists, we have discovered a series of dipyridodiazepinone inhibitors of HIV-1 RT polymerase (Merluzzi et al., 1990). A synthesis program based on potency for RT as well as a favorable metabolic and pharmacological profile has led to several potent RT antagonists. One compound, BI-RG-587 (nevirapine), has demonstrated a high degree of activity against HIV-1 RT and HIV-1 replication and was devoid of muscarinic and benzodiazepine activities. The synthesis and structure activity of the series of tricyclic diazepinones has been described in detail elsewhere (Hargrave et al., 1991).

RESULTS AND DISCUSSION

Effect of BI-RG-587 on HIV-1 rt and related enzymes

BI-RG-587 has an IC_{50} of 84 nM against HIV-1 RT but is not active against feline leukemia virus RT, simian immunodeficiency virus RT and HIV-2 RT. This

Innovations in Antiviral Development and the Detection of Virus Infection
Edited by T. Block *et al.*, Plenum Press, New York, 1992

compound inhibits HIV-1 RT using either homopolymer or heteropolymer (authentic HIV-1 genomic polymer) templates (Merluzzi et al., 1990; Kopp et al., 1991). The inhibition by BI-RG-587 is non-competitive with respect to both template-primer and nucleotide substrates (Kopp et al., 1991). The absence of an effect on K_m for either substrate indicates that BI-RG-587 interacts with HIV-1 RT at a site distinct from template-primer or nucleotide binding sites (Figure 1). The Ki for inhibition of HIV-1 RT polymerase is 220 nM and this compound has no effect on the processivity of the polymerase (Kopp et al., 1991). In addition, BI-RG-587 does not inhibit any mammalian DNA polymerase (alpha, beta, delta, or gamma) (Merluzzi et al., 1990; Kopp et al., 1991). Several other enzymes including HIV-1 protease and HSV-1 ribonucleotide reductase are also unaffected by this compound. BI-RG-587 has an IC_{50} of 50 nM against HIV-1 RNAse H, although complete inhibition (I_{max} = 54%) cannot be achieved. The partial inhibition of RNAse H may be due to the binding of BI-RG-587 at a site distinct from the active site (Merluzzi et al., 1990). These observations suggest that BI-RG-587 may cause a conformational change and may influence the HIV-1 RT polymerase site to a greater degree than the RNAse H site (Figure 1). The IC_{50} of BI-RG-587 for the HIV-1 RT DNA dependent DNA polymerase reaction is approximately 100 nM. The activity of BI-RG-587 against all three enzymatic activities of the HIV-1 RT molecule together with its specificity for this enzyme and not other RT enzymes and cellular polymerases indicates that this compound is very specific and targeted in action. A summary of the effects of BI-RG-587 on HIV-1 RT and other enzymes is shown in Table 1.

Molecular Target of BI-RG-587

An azido photoaffinity probe analog based on the structure of BI-RG-587, was found to irreversibly inhibit the HIV-1 RT enzyme reaction under UV irradiation (Wu et al., 1991). BI-RG-587 and analogs competitively protected HIV-1 RT from inactivation with the photoaffinity probe whereas substrates (dGTP, template-primer and tRNA) did not protect the enzyme. A tritiated photoaffinity probe selectively labelled the p66 polypeptide of HIV-1 at a 1:1 ratio (Wu et al., 1991).

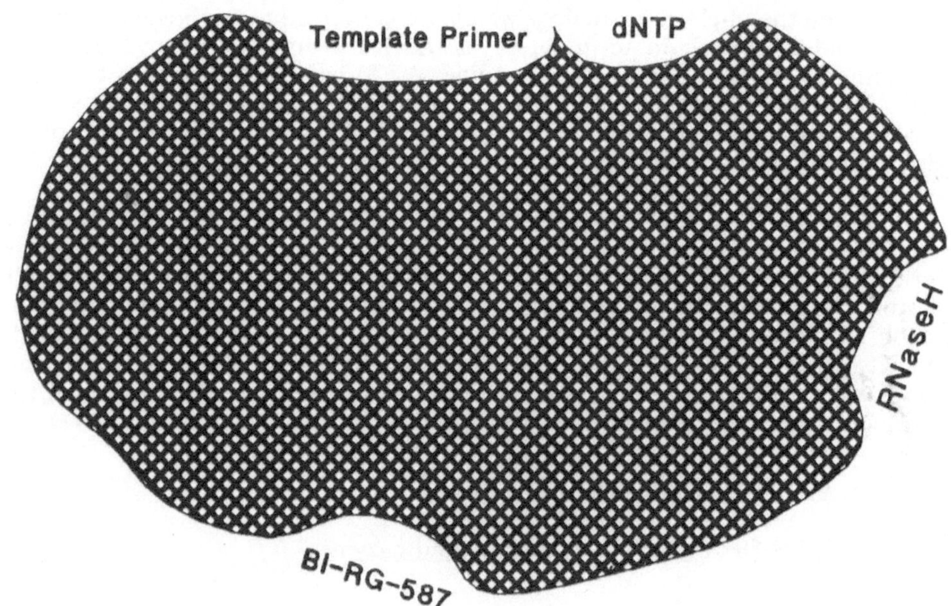

Figure 1. BI-RG-587 binds to RTase at a site distinct from template-primer or dNTP sites

Table 1. Effect of BI-RG-587 on HIV-1 RT and Enzyme Specificity

Enzyme[a]	Inhibition by BI-RG-587[b]
HIV-1 RNA-dependent DNA Polymerase	+
HIV-1 DNA-dependent DNA Polymerase	+
HIV-1 Ribonuclease H	+
HIV-1 Protease	-
HIV-2 RNA-dependent DNA Polymerase	-
FLV RNA-dependent DNA Polymerase	-
SIV RNA-dependent DNA Polymerase	-
Calf Thymus DNA-dependent DNA Polymerase-alpha	-
Human DNA-dependent DNA Polymerase-alpha	-
Human DNA-dependent DNA Polymerase-beta	-
Human DNA-dependent DNA Polymerase-gamma	-
Human DNA-dependent DNA Polymerase-delta	-
HSV-1 Ribonucleotide Reductase	-

[a] FLV: feline leukemia virus; SIV: simian immunodeficiency virus; HSV-1: herpes simplex virus type 1

[b] (+) indicates positive inhibition by BI-RG-587; (-) indicates no inhibition by BI-RG-587

Effect of BI-RG-587 on HIV Replication in Culture

BI-RG-587 inhibits HIV-1 replication in c8166 T cell cultures having IC_{50} against $HIV\text{-}1_{IIIb}$ of 40 nM with an I_{max} of 100% as determined by cytopathic effect (CPE) and 10 nM with an I_{max} of 100% as determined by inhibition of p24 production (Merluzzi et al., 1990; Koup et al., 1991). Viability in this cell culture system was determined by means of a tetrazolium salt (MTT) metabolic assay. This assay shows 50% cytotoxicity of BI-RG-587 at 321,000 nM providing a selectivity ratio in vitro of > 8000 (Merluzzi et al., 1990). Other isolates have been tested including the $HIV\text{-}1_{RF}$ (Haitian isolate) strain and isolates from patients. In all cases, BI-RG-587 was effective in reducing CPE and p24 production with similar IC_{50}'s (Merluzzi et al., 1990). Maximum inhibition was also achieved in all isolates and strains.

The effect of BI-RG-587 on HIV-2 was tested in a plaque reduction assay using CD4-transfected HeLa cells and a syncytia assay using c8166 T-lymphoblastoid cells. In both cases, this compound was inactive against HIV-2 (Merluzzi et al., 1990; Richman et. al., 1991). These data were not unexpected because this compound did not inhibit SIV RT and HIV-2 RT (Merluzzi et al., 1990). SIV and HIV-2 are more homologous than HIV-1 and HIV-2. *In situ* hybridization experiments using peripheral blood mononuclear cells (PBMC) and a clinical isolate of HIV-1 have shown that BI-RG-587 inhibits the accumulation of HIV-1 RNA. In the absence of BI-RG-587, 69/830 cells were positive for HIV-1 RNA. In the presence of BI-RG-587, 5/93,000 cells were positive for HIV-1 RNA (Merluzzi et al., 1990). These results were significant in that these isolates were never adapted to grow in cells other than fresh PBMC and they were tested in these experiments after their third passage (Merluzzi et al., 1990).

Since the target for RT inhibition by BI-RG-587 is different from that of nucleoside analogs which are chain terminators, we hypothesized that BI-RG-587 would be active against AZT resistant strains of HIV-1 and may possibly synergize with AZT for inhibition of HIV-1 replication in culture. In fact, these two hypotheses are correct. BI-RG-587 was effective against AZT resistant isolates. The isolates were pre-selected and post-selected from patients undergoing therapy with AZT. The post-selected isolates had a several-fold decrease in sensitivity of AZT while BI-RG-587 was as effective on AZT resistant isolates as on those pre-selected (AZT sensitive) (Richman et al., 1991). BI-RG-587 was also synergistic with AZT (Richman et al., 1991). Other studies have shown that BI-RG-587 inhibits HIV-1 p24 production in CEM T-lymphoblastoid cells and in primary human monocyte cultures (Richman et al., 1991). This compound did not inhibit cytopathogenicity in HeLa cells infected with rhinovirus 54 or poliovirus I, influenza A virus induced CPE on bovine kidney cells or vaccinia virus induced CPE on human lymphoid cells (Merluzzi et al., 1990; Koup et al., 1991; Richman et al., 1991). These results show further specificity of BI-RG-587 in its inability to inhibit RNA and DNA viruses not dependent upon reverse transcriptase in their life-cycle. A summary of the effects of BI-RG-587 on virus replication is shown in Table 2.

Immunology, Pharmacology and Pharmacokinetics of BI-RG-587

Preliminary studies indicate that BI-RG-587 is not immunosuppressive and is not toxic to bone marrow progenitors (Merluzzi et. al., 1991). In addition, initial studies of metabolic and tissue distribution studies were carried out in rodents and primates. Such tissue distribution studies after oral administration indicate a plasma:brain ratio of 0.8 to 1.0. These results are encouraging for the development of BI-RG-587 and its analogs as antiviral agents for the treatment of HIV-1 infection in humans. The non-nucleoside nature of this compound may circumvent the toxicities normally associated with nucleoside chain-terminators of reverse transcriptase.

Table 2. Effects of BI-RG-587 on Virus Replication

Virus[a]	Target Cell[b]	Measurement[c]	Inhibition by BI-RG-587[d]
HIV-1_{IIIb}	c8166 T-cells	Syncytia	+
HIV-1_{IIIb}	c8166 T-cells	p24	+
HIV-1_{IIIb}	Human PBMC	p24	+
HIV-1_{RF}	c8166 T-cells	Syncytia	+
HIV-1_{UMGL}	c8166 T-cells	Syncytia	+
HIV-1 Isolates	Human PBMC	p24	+
LAV-1_{BRU}	CD4+ HeLa	Plaque	+
LAV-1_{BRU}	CEM T-cells	p24	+
HTLV-III_{Bal-85}	Human monocytes	p24	+
HIV-1 AZTR Isolates	CD4+ HeLa	Plaque	+
HIV-2_{ROD}	CD4+ HeLa	Plaque	-
HIV-2_{ROD}	c8166 T-cells	Syncytia	-
Rhinovirus 54	HeLa	CPE	-
Coxsackie A13	HeLa	CPE	-
Influenza A	MDBK	CPE	-
Poliovirus type I	HeLa	CPE	-

[a] AZTR: AZT-resistant clinical isolates.
[b] PBMC: Peripheral Blood Mononuclear cells; MDBK: Bovine Kidney Cells
[c] CPE: cytopathic effect
[d] (+): Inhibition of virus replication; (-): No inhibition of virus replication.

REFERENCES

Gilboa, E., Mitra, S.W., Goff, S., and Baltimore, D., 1979, A detailed model of reverse transcription and tests of crucial aspects, Cell, 18:93-100.

Ho, D.D., Moudgil, T., and Alam, M., 1989, Quantitation of human immunodeficiency virus type I in the blood of infected persons, New. Eng. J. of Med., 321:1621-1625.

Hargrave, K.D., Hehnke, M.L., Cullen, E., Engel, W.W., Fuchs, V., Grozinger, K.G., Kapadia, S.R., Klunder, J.M., Mauldin, S., McNeil, D., Pal, K., Patel, U., Proudfoot, J.R., Rose, J., Schmidt, G., Skiles, J.W., Skoog, M., Vitous, J., and Adams, J., 1991, Novel Non-nucleoside inhibitors of HIV-1 reverse transcriptase. I. Tricyclic Benzopyrido- and Dipyridodiazepinones, J. Med. Chem., 34:2231-2241.

Kopp, E.B., Miglietta, J.J., Shrutkowski, A.G., Shih, C.K., Grob, P.M., and Skoog, M.T., 1991, Steady state kinetics and inhibition of HIV-1 reverse transcriptase by a non-nucleoside dipyridodiazepinone, BI-RG-587, using a heteropolymeric template, Unpublished observations.

Koup, R.A., Merluzzi, V.J., Hargrave, K.D., Adams, J., Grozinger, K., Eckner, R.J., and Sullivan, J.L., 1991, Inhibition of HIV-1 replication by the dipyridodiazepinone, BI-RG-587, J. Infec. Dis., 163:966-970.

Larder, B.A., Darby, G., and Richman, D.D., 1989, HIV with reduced sensitivity to Zidovudine (AZT) isolated during prolonged therapy, Science, 243:1731-1734.

Merluzzi, V.J., Hargrave, K.D., Labadia, M., Grozinger, K., Skoog, M., Wu, J., Shih, Cheng-Kon, Eckner, K., Hattox, S., Adams, J., Rosenthal, A.S., Faanes, R., Eckner, R.J., Koup, R.A., and Sullivan, J.L., 1990, Inhibition of HIV-1 replication by a non-nucleoside reverse transcriptase inhibitor, Science, 250:1411-1413.

Merluzzi, V.J., Faanes, R.B., Richman, D., and Moore, M.A.S., Comparison of BI-RG-587 and Zidovudine on Human Bone Marrow Progenitors, 1991, Unpublished observations.

Mitsuya, H., Weinhold, K.J., Furman, P.A., St. Clair, M.H., Nusinoff-Lehrman, S., Gallo, R.C., Bolognesi, D., Barry, D.W., and Broder, S., 1985, 3'Azido-3'deoxythymidine (BWA509U): An antiviral agent that inhibits the infectivity and cytopathic effect of human T-lymphocyte virus type III/lymphadenopathy-associated virus *in vitro*, Proc. Natl. Acad. Sci., 82:7096-7100.

Richman, D.D., Fischl, M.A., Grieco, M.H., Gottlieb, M.S., Volberding, P.A., Laskin, O.L., Leedom, J.M., Groopman, J.E., Mildvan, D., Hirsch, M.S., Jackson, G.G., Durack, D.T., Nusinoff-Lehrman, S., and The AZT Collaborative Working Group, 1987, The toxicity of azidothymidine (AZT)in the treatment of patients with AIDS and AIDS-related complex, New Eng. J. of Med., 317:192-197.

Richman, D., Rosenthal, A.S., Skoog, M., Eckner, R.J., Chou, T.C., Sabo, J.P., and Merluzzi, V.J., 1991, BI-RG-587 is active against Zidovudine-resistant human immunodeficiency virus type 1 and synergistic with Zidovudine, Antimicrob. Agents and Chemother., 35:305-308.

Wu, J.C., Warren, T.C., Adams, J., Proudfoot, J., Skiles, J., Raghavan, P., Perry, C., Potocki, I., Farina, P.R., and Grob, P.M., 1991, A novel dipyridodiazepinone inhibitor of HIV-1 reverse transcriptase acts through a non-substrate binding site, Biochemistry, 30:2022-2026.

Yarchoan, R., Klecker, R.W., Weinhold, K.J., Markham, P.D., Lyerly, H.K., Durack, D.T., Gelman, E., Lehrman, S.N., Blum, R.M., Barry, D.W., Shearer, G.M., Fischl, M., Mitsuya, H., Gallo, R.C., Collins, J., Bolognesi, D., Myers, C., and Broder, S., 1986, Administration of 3'-azido-3'deoxy-thymidine, an inhibitor of HTLV-III/LAV replication, to patients with AIDS or AIDS-related complex, Lancet, 1:575-580.

CATALYTIC ANTISENSE RNA (RIBOZYMES): THEIR POTENTIAL AND USE AS ANTI-HIV-1 THERAPEUTIC AGENTS

John J. Rossi[1], and Nava Sarver[2]

[1]Department of Molecular Genetics, Beckman Research Institute of the City of Hope, Duarte, CA. 91010

[2]Developmental Therapeutics Branch
Division of AIDS, NIAID
Bethesda, MD 20892

INTRODUCTION

Catalytic RNAs were first discovered as natural processes in several biological systems [1,2,3]. Their discovery is one of the landmark accomplishments of modern science since it was the first demonstration that an informational molecule could also possess enzymatic activity, and hence changed our views on the possible origins of life. The discovery and characterization of the reactions carried out by RNA molecules has facilitated the development of an entirely distinct approach to antiviral therapy, the site specific, cleavage of viral RNAs mediated by catalytic, anti-sense ribozymes. We are using ribozymes as a novel antiviral strategy. In the following paragraphs, we weill describe the development of anti-sense ribozymes with the primary emphasis being their application as anti-HIV-1 therapeutic agents.

The relatively simple self-cleavage domain of certain plant viroids and virusoids, termed the "hammerhead"[3] has lent itself to the development of sequence specific catalysts which facilitate self-cleavage of target RNAs. This bi-molecular cleavage domain is utilized during the replication of these plant pathogens as described in Figure 1. First, there is an infection by the RNA in the form of a viroid or co-encapsulation with a plant virus (virusoid) and the input of a positive strand viral genome. The positive strand serves as a template for negative strand synthesis via a rolling circle mechanism generating a multinumeric RNA product which then breaks down to monomeric units by self-cleavage at a specific site in the RNA. The specific self-cleavage site is generated by folding of the RNA upon itself. The unit length monomers generated are then circularized, most likely by a plant ligase. These in turn serve as templates for the formation of an oligomer of positive polarity. The positive multimer folds upon itself, creating the hammerhead catalytic center which allows self-cleavage to unit length molecules of positive polarity. These monomers are recircularized as described above, regenerating the infectious positive strand RNAs. This catalytic process does not require an exogenous energy source, proteins or cofactors. This is one of the considerations making this strategy adaptable for therapeutic applications.

Innovations in Antiviral Development and the Detection of Virus Infection
Edited by T. Block *et al.*, Plenum Press, New York, 1992

1. Rolling circle replication of infectious plus strand.

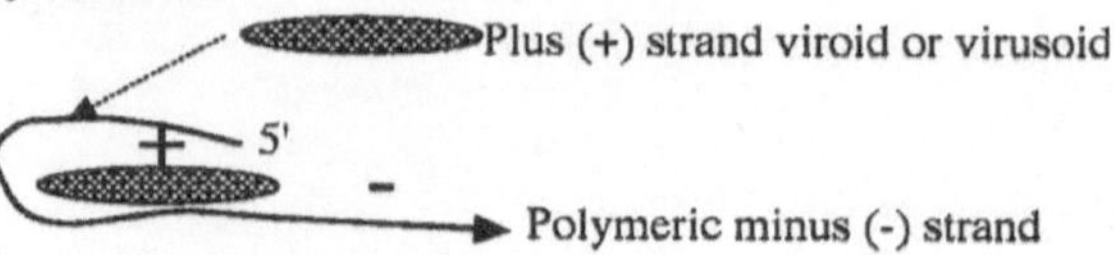

2. Autocleavage into monomers.

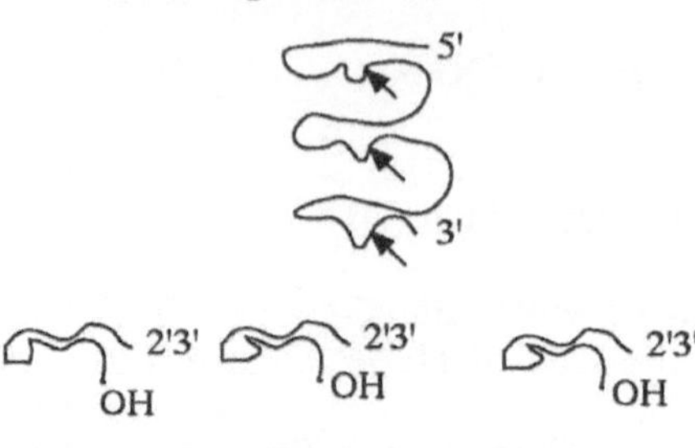

3. Recircularization of minus strands.

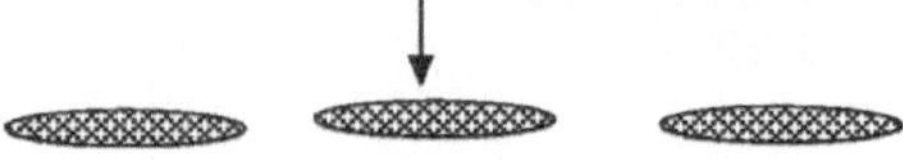

4. Cycle is repeated for formation of plus strands.

Figure 1 Plant viroid and virusoid replication and autocleavage cycle involving hammerhead autocleavage domain. This diagram was adapted from materials presented in Riesner and Gross[15].

HAMMERHEAD AUTOCLEAVAGE DOMAIN

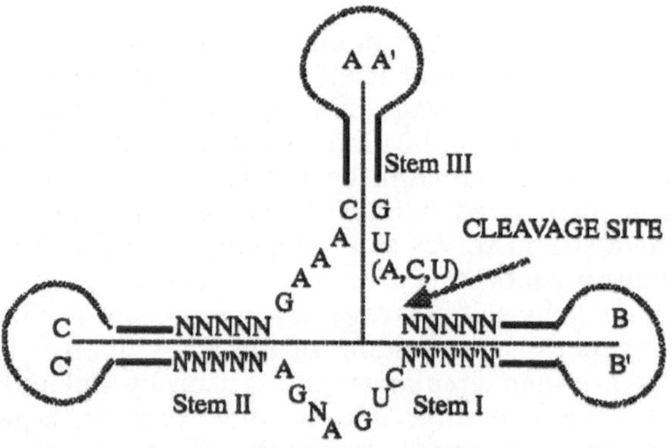

Figure 2. A schematic diagram of hammerhead autocleavage domain, one of the self-catalytic motifs identified in plant viroids and virusoids. The dotted lines delineate axes which separate functional units of the hammerhead for application in *trans*- catalysis reactions.

One of the catalytic regions associated with the self-cleavage process, termed the "hammerhead" domain is shown in Figure 2. There are 13 highly conserved nucleotides illustrated in this figure, which are present in the plant viroid, virusoid, and a newt autocleavage domain. [3,4]. The stems I,II, and III are formed by base pairing, and are critical determinants in the formation of the catalytic active center. Their precise sequence is not critical as long as duplex formation is retained. Recent data, however, suggest that the primary sequence may influence the rate of the reaction[5]. Uhlenbeck first demonstrated that the catalytic active center could be created by an intermolecular pairing of two separate RNA strands.[6]. This simple ribozyme consisted of two oligomeric RNAs brought together by base pairing as illustrated by C-B and C'-B' in Figure 2. The lower strand (ribozyme) catalyzes a cleavage of the upper strand (substrate) in the presence of magnesium ions. This process is catalytic, and multiple rounds of annealing, cleavage and product dissociation have been demonstrated. (see Figure 3).

We are looking at ribozymes as potential therapeutic agents for the treatment of HIV-1 infection. Ribozymes can be considered as an enzymatic form of anti-sense, capable of specific base pairing to a target RNA, followed by cleavage of the phosphodiester backbone of that target. The hammerhead motif presents an excellent model for the applications of this technology for the following reasons. First, the sequence required for effecting cleavage is known, and ribozymes targeted to specific RNA substrates can be engineered, synthesized and readily tested in the laboratory. Analogs, perhaps with enhanced activity, may also be designed. Second, approaches are available to stabilize the ribozyme within a cell, without compromising the catalytic activity. Third, methods are being designed to deliver any type of nucleotide, be it DNA or RNA, into cells, thus providing vehicles for the delivery of ribozymes as either a pre-formed drug, or in the form of a gene for constitutive expression (gene delivery). Fourth, there are systems available for evaluating the effectiveness of ribozymes, making it possible to test ribozymes for their anti-viral activity.

Figure 3 illustrates a typical pathway for cell free cleavage. The reaction is absolutely dependent upon magnesium. The hammerhead cleavage domain depicted in Figure 2 allows for the prediction of three different *trans* interactions. The first is bisected by lines C-B' and C'-B'. In this model, the cleavage site is in the upper strand (substrate) while the lower strand serves as the catalyst. The other interactions are defined by A'-B with A-B' and a tripartite interaction defined by interactions of A'-B, A-C and C'-B'. In the A'-B and A-B' interactions, which were discovered by Haseloff and Gerlach[7] , the majority of the catalytic center can be supplied in *trans*, thereby allowing for great flexibility in the choice of target site. Shown in this figure, the cleavage site is defined by the sequence GU, followed by A, C, or U. It has been demonstrated that any XU, A, C, or U site can be cleaved, but there is clearly a preference for GUC, and XUG is an extremely poor cleavage substrate.[8]

ANTI-HIV-1 RIBOZYMES

Two *trans*-acting hammerhead designs targeted to the same HIV-1 *gag* cleavage site were compared for cleavage activity as depicted in Figure 4. Design A has an equal distribution of the conserved domains, such that the HIV-1 target sequence supplies one half of the catalytic center. In B, based upon the work of Hasseloff and Gerlach[7] the substrate only provides the GUC at the cleavage site, and the catalyst, supplied in *trans*, provides the remaining catalytic domains. Cleavage reactions in which stoichiometric amounts of catalyst and substrate were incubated together are depicted. From the results presented here, two major conclusions can be drawn. First, HIV-1 RNA can constitute a substantial portion of a hammerhead ribozyme and participate in a self-cleavage reaction as exemplified by the ribozyme substrate interactions in

Model A. Second, the B design is much more active under the conditions chosen, with complete cleavage of the target molecule. Since the B design allows much greater flexibility in target selection (the only limitation is the occurrence of XU, followed by A, C, or U), this is the design of choice for most applications.

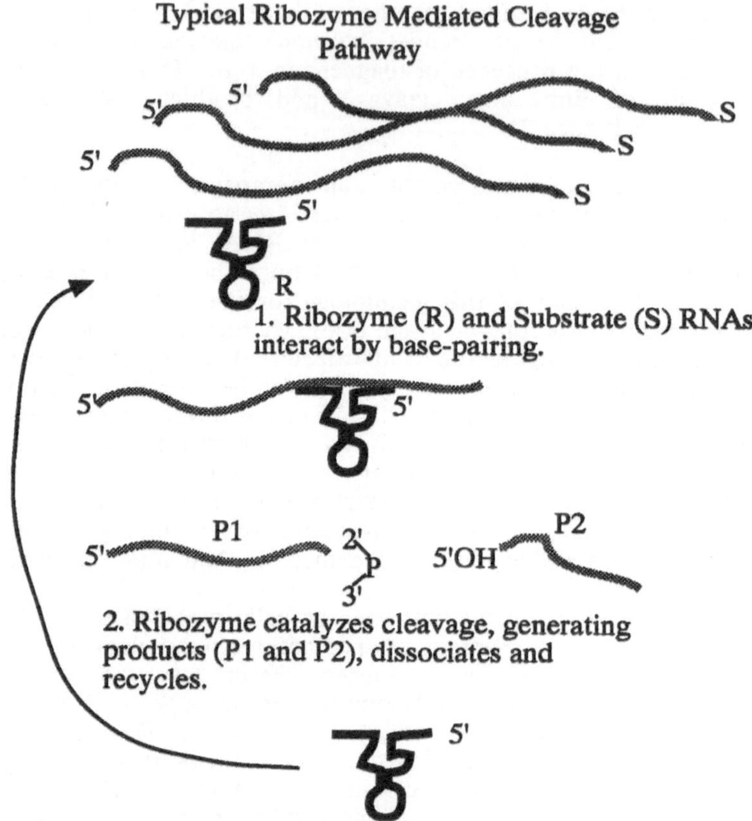

Figure 3. Diagrammatic representation of *trans*-acting hammerhead ribozyme mediated cleavage of a target RNA. The ribozyme substrate interaction is via base-pairing. The cleavage reaction occurs only in the presence of magnesium ions.

A titration experiment of increasing substrate concentration relative to a fixed catalyst concentration is depicted in Figure 5 in which the ratio of substrate to catalyst increases from 1:1 to 50:1. The important conclusion from this experiment is that there is greater than stoichiometric cleavage of the substrate at all ratios of substrate to catalyst greater than 1. Thus, under near physiological temperature and salt conditions, a ribozyme catalyst such as the one depicted here can turn over multiple substrate molecules.

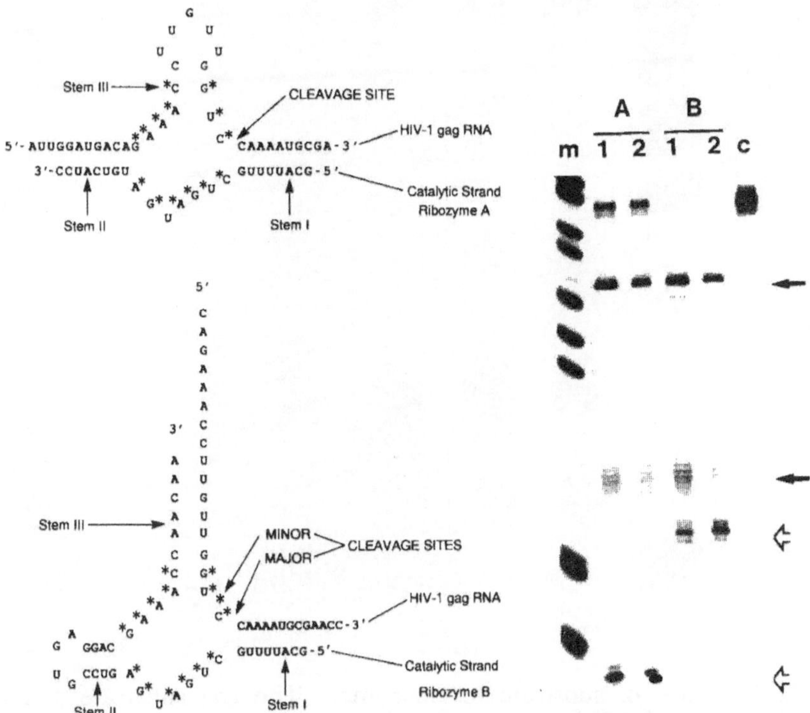

Figure 4. Comparative cleavage reactions of two different models for hammerhead ribozymes. Both ribozymes are targeted to the same HIV-1 *gag* RNA target described in references 9 and 10. Stems I, II, and III correspond to those depicted in Figure 2. The upper ribozyme reaction is represented in Panel A, lanes 1 and 2, while the lower ribozyme reaction is represented in Panel B, lanes 1 and 2. The difference between lanes 1 and 2 in each case is that the ribozyme was capped with GpppG in lane 1, and uncapped in lane 2. The ratio of ribozyme to substrate is 1:1 in each case, and the time course of cleavage is 14 hours at 37°C. Details of this experiment are described in Chang, *et al*[9]. The upper solid arrows point to the 5' (upper) and 3' (lower) cleavage products. The open arrows point to the ribozyme A or B. The lane C, is a control *gag* RNA substrate incubated under cleavage conditions in the absence of added ribozyme RNAs.

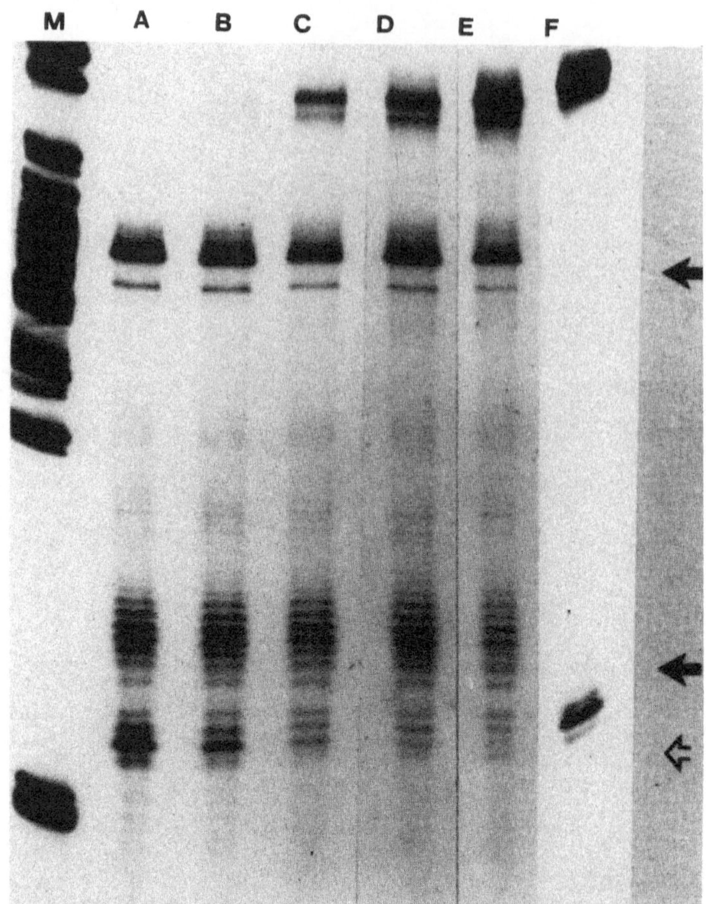

Figure 5. Titration of substrate to ribozyme. The *gag* substrate and ribozyme B depicted in Figure 4 were used in this cleavage reaction. The titration of substrate to ribozyme is as follows: A, 1:1, B, 5:1, C, 10:1, D, 25:1, and E, 50:1. Lane F is a control in which a 1:1 substrate to ribozyme ratio was incubated under identical conditions to the other lanes, except magnesium was omitted from the reaction. The conditions for cleavage were 14 hours at 37°C in 20mM magnesium.

SPECIFICITY AND CLEAVAGE IN A COMPLEX MILIEUX

An important question concerning the use of ribozymes as antiviral therapeutics is that of specificity of cleavage in a complex RNA environment. This question was tested in the following experiment (described in Chang *et al.*,[9] and Rossi, *et al.*,[10]). An anti-HIV-1 *gag* ribozyme was incubated with HIV-1 *gag* RNA transcribed in a cell free system. Included in the reaction mixture were increasing concentrations of total RNA prepared from either HIV-1 infected or uninfected, human T lymphocytes. The rationale of this experiment is as follows. If the ribozyme is interacting with non-targeted RNAs, increasing amounts of total RNA from uninfected lymphocytes should result in reduced cleavage of the radioactively labelled *gag* substrate. If non-specific interactions are insignificant, increasing concentrations of total RNA from HIV-1 infected cells should have inhibitory effects on the cleavage efficiency of the labelled *gag* target because of competing *gag* sequences present in total RNA obtained from infected cells. Indeed, increasing amounts of total cellular RNA from uninfected cells had no effect on the cleavage reaction whereas RNA from HIV-1 infected cells inhibited cleavage of the labelled *gag* substrate. When the ribozyme concentration was increased, the inhibitory effect was overcome. We conclude from these experiments that the specific cleavage activity of the ribozyme was unaffected by non-targeted sequences, and that the RNA from virally infected cells was competing with the synthetic substrate. Taken together, these data suggest that ribozymes may be capable of specific cleavage reactions in an intracellular environment.

CONSTITUTIVELY EXPRESSED ANTI-HIV-1 RIBOZYMES CAN RENDER CELLS IMMUNE TO HIV-1 INFECTION

To test the effectiveness of an anti-HIV-1 ribozyme in an intracellular environment the following approach was taken: mammalian cells expressing the CD4 receptor were transfected with a vector harboring an anti-HIV-1 *gag* ribozyme gene driven by the human β-actin promoter. The constitutive expression of that ribozyme within the cell should protect the cells against HIV infection by cleaving incoming viral RNAs and the *gag* mRNAs produced during the course of viral infection. To address the question of whether ribozymes expressed constitutively are biologically active, total RNA was extracted from transfected cells and reacted with *gag* substrate prepared *in vitro*. The results are presented in Figure 6a. In lane "B", the catalyst and substrate were synthetic; the location of one of the two cleavage products is indicated for reference. The second ribozyme cleavage product is small, and not detected in this gel system. In lanes C, D and E the reactions were carried out with decreasing amounts (5, 1 and 0.2 μgm) of total RNA extracted from transfected cells and was reacted with synthetic substrate. Although the cleavage product is not as prominent as that obtained with the synthetic ribozyme, small amounts of the cleavage products were detected which comigrate with the band generated with the synthetic ribozyme. This band is not present with 5 μgm of total RNA extracted from ribozyme negative cells (lane A). These experiments demonstrated that the ribozyme produced by transcription from the β-actin promoter is catalytically active, despite the fact that it harbors a large excess of non-complementary sequence derived from the β-actin 5' untranslated region, and the SV-40 sequences adjacent to the polyadenylation signal, and a stretch of poly-Adenosine.

A major concern of any type of antiviral strategy is potential cytotoxicity. In a simple experiment where the growth curve of cells constitutively expressing the anti-HIV-1 ribozyme was followed over a period of 10 days, no impairment of cell growth was observed (Figure 6b). Cytotoxicity studies with longer term cultures also have shown that ribozyme containing cells behave the same as their control, non-ribozyme expressing counterparts.

Cells which constitutively express the anti-HIV-1 *gag* ribozyme were challenged with HIV-1. The ribozyme expressing clones had a marked reduction in soluble p24 ($>$50 fold), one of the antigen markers for viral load. In the same experiment, parental cells infected with HIV-1 had $>$10 ng per ml of p24 present in the conditioned medium[11]. Eight different isolates of protected cells show various reduction in p24, down to 0.23 and 0.07 μgms per ml of secreted p24. Since p24 is one of the obligatory proteins for virus assembly, these data strongly argue that a ribozyme can impact on the replication of the virus and progeny production from these infected cells.

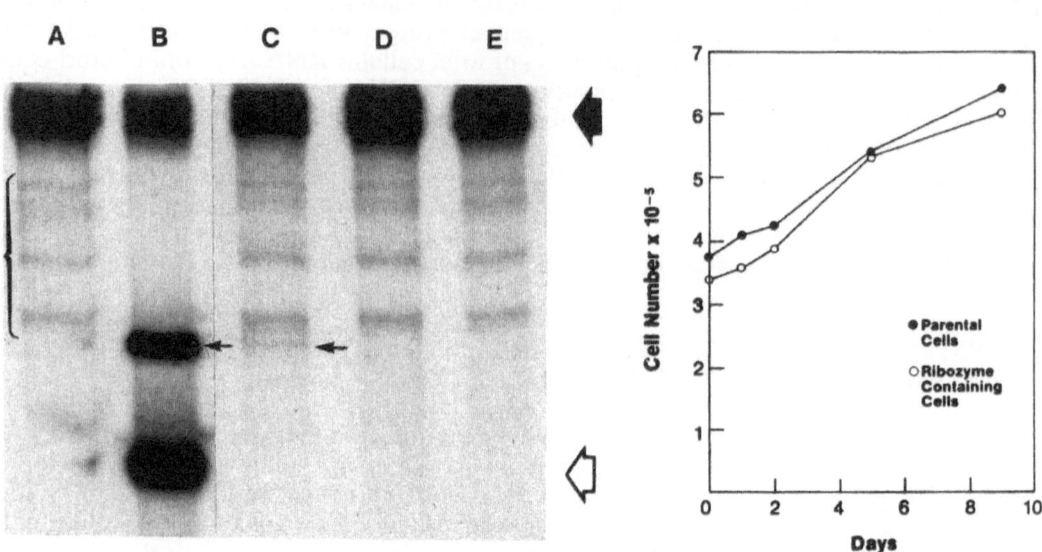

Figure 6. Cleavage of *gag* substrate RNA by intracellularly expressed ribozyme (a) and growth comparisons of ribozyme expressing and parental CD4+ Hela cells (b). a. *In vitro* cleavage of HIV-1 *gag* substrate by ribozyme produced from transcription in cell culture. Lane A, 5 μgms total RNA from parental Hela cells incubated with "*gag*" substrate; Lane B, substrate plus *in vitro* transcribed anti-*gag* ribozyme at 1:1 molar ratios; Lane C, substrate incubated with 5 μgms of total RNA from CD4+ Hela cells expressing anti-*gag* ribozyme; Lane D, substrate incubated with 1 μgm of total RNA from ribozyme expressing CD4+ Hela cells; Lane E, substrate incubated with 0.2 μgms of total RNA from ribozyme expressing cells. The heavy arrow points to the *gag* substrate RNA, the smaller arrow points to the 5' cleavage product. The smaller 3' cleavage product was electrophoresed off of this gel. For details, see Chang, *et al.*[9] b. Growth of ribozyme expressing versus parental CD4+ Hela cells. Cells were cultured for the indicated time period and counted on days 1, 3, 5, and 9.

To investigate further whether the ribozyme had impacted on HIV-1 viral RNA, we carried out the following experiment. This involved a polymerase chain reaction (PCR)[12] assay utilizing a primer set which flanks the ribozyme cleavage site in the HIV-1 *gag* gene at nucleotides 805-806. The rationale is that the amount of amplified product generated from ribozyme protected cells should be markedly reduced in relation to non-protected cells since the priming is designed to generate products which traverse the cleavage site. As a control, a primer set designed to amplify sequences downstream of the cleavage site was used. The results presented in Figure 7 demonstrate a dramatic reduction in the PCR product from the primers traversing the cleavage site (480 nt product). The 200 nt PCR product derived from sequences downstream of the cleavage site is also somewhat reduced relative to the non-ribozyme expressing parental cells, but clearly not as much as the 480 nt product. One interpretation of these results is that the constitutively expressed ribozyme has cleaved the HIV-1 RNA such that only small amounts of 480 nt product can be generated.

The less pronounced reduction in 200 nt product can be attributed to an overall reduction in viral genomes in this cell population due to cleavage by the ribozyme. These data taken together with the antigen data provide strong support to the conclusion that a ribozyme can afford marked protection to HIV-1 infection.

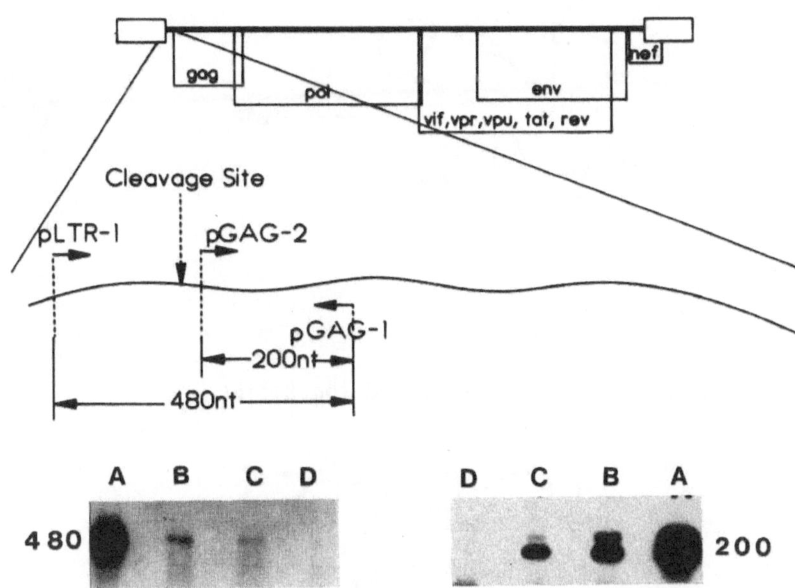

Figure 7. PCR assay for ribozyme mediated cleavage using total RNA from HIV-1 infected cells. The top diagram depicts the region of the *gag* gene targeted by the ribozyme. The positions of the primers used in the RNA based PCR reaction are indicated. The bottom panels depict probe hybridizations to the PCR amplified products. The probe used was common to both amplified products. Lane A is amplified DNA from the parental CD4+ Hela parental cell line; Lane B is from a single ribozyme expressing clone; lane C is from a pool of several ribozyme expressing clones; lane D is a contamination control. For details, see Sarver, *et al.*,[11].

ADVANTAGES OF RIBOZYMES VERSUS CONVENTIONAL ANTISENSE

What are the advantages of a ribozyme vis-a-vis conventional antisense? First, the cleavage of the virus genome or subgenomic mRNAs is one of the primary benefits that can accrue with ribozymes. Not only do they neutralize the gene but they also cleave it, guaranteeing its permanent inactivation. Secondly, the process is catalytic, and, therefore, at least theoretically, lower concentrations are required than conventional antisense RNAs. A third advantage is that of targeting, best illustrated in Figure 8. The standard antisense approach involves long RNAs having their own inherent structural properties which may in fact preclude the two molecules from interacting. This problem can be somewhat circumvented by small antisense RNAs, but these may not be very stable, and since they only bind to their target, may be easily displaced by translation or other protein interactions.

1. **RNAs have inherent structure and may not pair.**

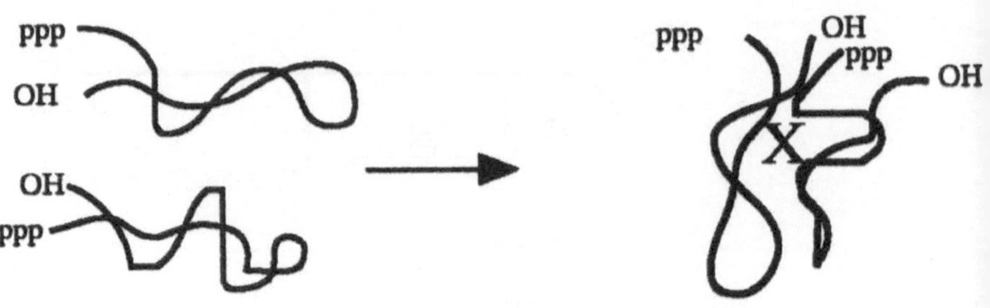

2. **Short antisense RNAs may be easily displaced from hybrid.**

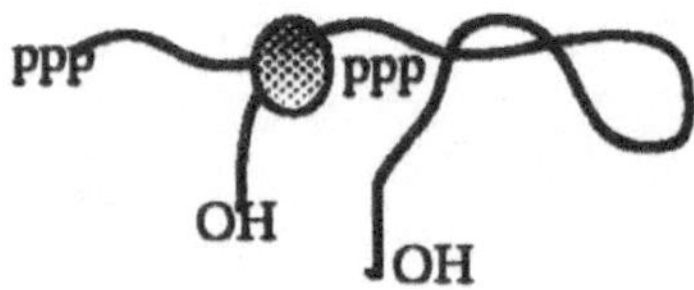

3. **Ribozymes have preformed structure, once paired with their substrate, they quickly cleave it.**

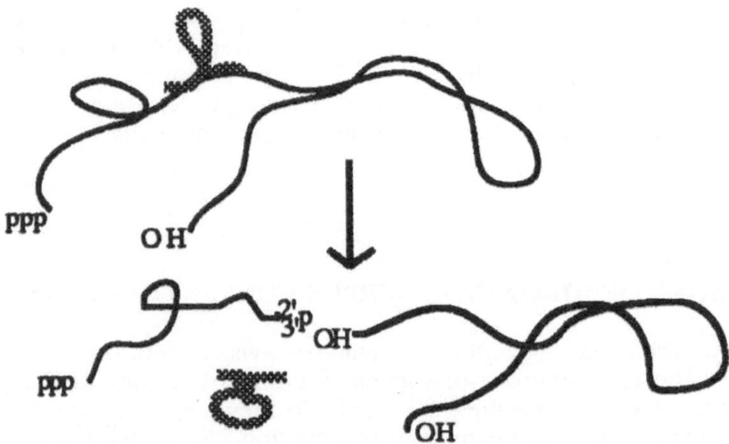

Figure 8. Target accessibility advantage of a ribozyme versus conventional antisense RNAs.

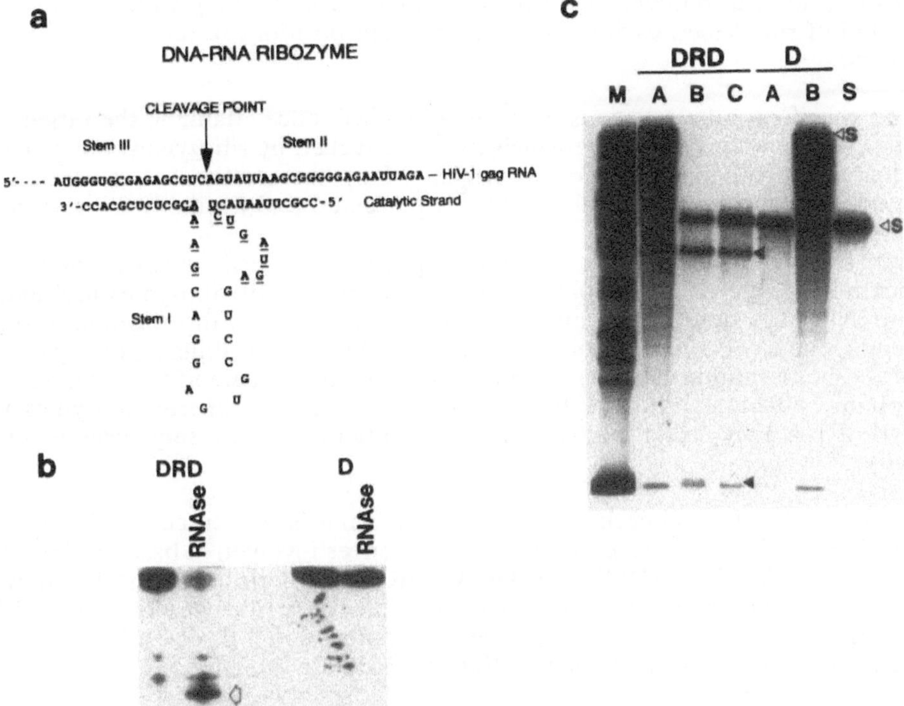

Figure 9. Cleavage by a chimeric DNA-RNA ribozyme targeted to HIV-1 *gag*. In panel a, the sequence of a conventional ribozyme and substrate target are indicated. The chimeric ribozyme has DNA at all positions except those which are underlined. In panel b, the RNAse sensitivity of the DRD ribozyme versus an all DNA (D) isomer are tested. The open arrow indicates the radioactive 5' end of the DRD molecule which is cleaved after the first underlined C by RNAse A. Panel C depicts a cleavage reaction mediated by the DRD ribozyme. The lanes are: M, molecular weight marker; under the DRD, A, DRD cleavage of a 610 nt *gag* target RNA; B and C, cleavage of a 240 nt *gag* transcript by two different DRD preparations; under the D, all DNA isomer incubated with the 240, lane A, or 610, lane B *gag* substrate; lane S is the 240 nt substrate alone. The letter S depicts the precursor substrates, and the carats indicate the 5' and 3' cleavage products of the 240 nt substrate. Because of breakdown, the large 610 nt cleavage product is not discernible in lane DRD-A.

One potential problem with a ribozyme therapy for HIV is that this virus has an apparently high capacity for genetic variability. To circumvent this problem, cleavage sites in the viral genome may be chosen which if mutated, render the virus non-functional. These include splice signals, packaging signals, and some of the cis-acting regulatory elements such as TAR and RRE. It is also possible to simultaneously target multiple sites along the viral genome. Since hammerhead type ribozymes are small molecules, a multivalent ribozyme capable of interacting with several sites, or a cocktail of ribozymes, each targeted to different positions in the viral RNA, can be employed.

The specificity of cleavage is contributed by sequences flanking the target. Since all diseases have an RNA stage which can be attacked by ribozymes, independent of whether they are cellular or viral in origin. Many human diseases can therefore be targeted with ribozymes once the system is perfected and available for clinical use.

With any new modalities there are problems which hamper their clinical application. Some of the problems confronting the use of ribozymes are addressed below. Nuclease degradation may be a major hurdle for both exogenously supplied and endogenously expressed ribozymes. Potential ways of stabilizing ribozymes are 5' capping, 3' end modifications, incorporation of modified nucleotides or altered sugar-phosphate backbones into the flanking sequences, and chimeric ribozymes with a catalytic RNA core and DNA (diester or modified) flanking sequences to increase stability.

An example of a chimeric RNA/DNA ribozyme is presented in Figure 9. The catalytic center is RNA, while the flanking sequences involved in base pairing with the target are DNA. This DNA-RNA-DNA (DRD) chimeric ribozyme is catalytically active, although its kinetics are slower than an all RNA ribozyme (Taylor and Rossi, manuscript in preparation). As might be anticipated, an all DNA molecule harboring the same sequences as the chimeric molecule is totally inactive.

Delivery of Ribozymes

Liposomes are being tested as a possible vehicle for the delivery of ribozymes to cells in a clinical setting. Many possible formulations for liposomes exist, and choosing the best one will depend on the target cells, the intracellular compartmentalization of the liposome formulation, and the loading capacity of the formulation[13]. We have been exploring liposome mediated delivery of ribozymes in cell culture. An example of a northern hybridization of ribozyme RNA delivered to human T4 lymphocytes by liposomes is presented in Figure 10. In comparing the liposome encapsulated versus unencapsulated RNA, it is clear that there is a significant intracellular accumulation of the liposome encapsulated ribozymes, whereas unencapsulated ribozyme RNA is not detected. At the time of this writing, it has not been demonstrated that liposome delivered ribozymes are protective against HIV-1 infection. Such experiments will provide information regarding the potential utility of liposomes for delivery of synthetic anti-HIV-1 ribozymes.

Our ultimate goal is to combine ribozymes with a gene therapy approach. Based upon recent human gene therapy experiments at the NIH and elsewhere, it is becoming increasingly apparent that gene therapy may be used to correct genetic deficiencies as well as to combat viral and bacterial pathogens. Several delivery strategies are being examined. Perhaps the most advanced is the use of retroviral vectors for hematopoietic delivery of genes. Other infectious, but not replicating vectors that can be used to deliver ribozyme gene constructs are also under evaluation.

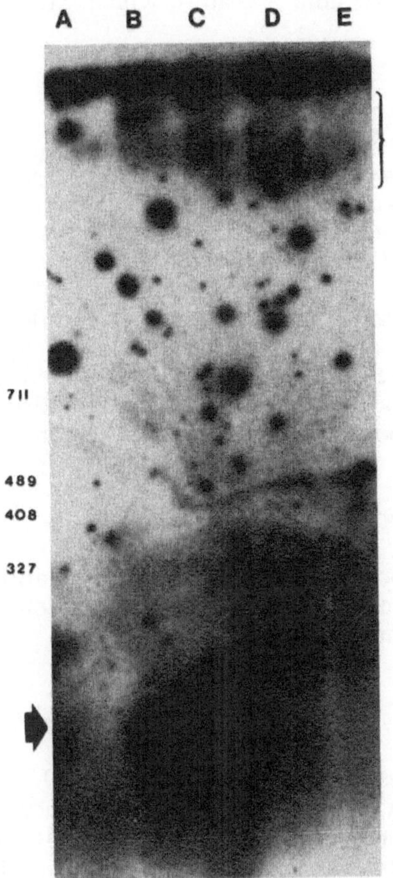

Figure 10. Northern gel analysis of anti-HIV-1 *gag* ribozyme delivered to H9 lymphocytes by liposomes. Following various times of uptake, total RNA from H9 cells was extracted, electrophoresed, blotted to a nylon membrane and hybridized to a radioactive probe complementary to an anti-HIV-1 *gag* ribozyme. In Lane A, the RNA was from untreated cells, in Lane B, from ribozyme encapsulated liposome treated cells allowed to uptake for 24 hours and in Lane C for 48 hours. Lanes D and E represent total RNA extracted from cells incubated with an equivalent amount of unencapsulated (naked) ribozyme for one hour (D) and 3 hours (E). The arrow points to the ribozyme sized band hybridized to the ribozyme complementary probe. The numbers are from a molecular weight marker run on the same gel.

Adeno-associated (AAV) viral vectors are very promising as gene therapy vehicles. AAV can be engineered to harbor 2-3 kb of foreign DNA, and efficient packaging and transfection systems have been developed[14]. One feature of this vector system which makes it worthy of consideration is that it appears to integrate non-randomly into a specific site in the long arm of chromosome 19[16]. Integrated constructs appear to maintain long-term expression, a feature which will be required for effective gene therapy.

CONCLUDING REMARKS

In summary, ribozymes are a novel form of anti-viral agent that draw upon state -of-the-art technologies in molecular genetics, chemistry, cell biology, virology and medicine for full realization of their potential. It is the intent of this article to convey some of the potential of ribozymes as well as some of the problems confronting their use-as anti-viral agents. It should be emphasized that ribozymes are simply an advanced form of anti-sense molecules, and as the fields of antisense and ribozymes progress in successful applications, there will be a concomitant increase in the successful use of these technologies. It is hoped that some of the important distinctions between ribozymes and conventional antisense have been made clear and advantages to using ribozymes as anti-viral agents are evident.

ACKNOWLEDGEMENTS

This work was supported in full or in part by National Institute of Health grants AI29329 and AI25959 to J.J.R. The authors are grateful to W.J. Mutter for the photo-ready preparation of this manuscript.

REFERENCE

1. Kruger, K., Grabowski, P.J., Zaug, A.J., Sands, J., Gottschling, D.E., Cech, T.R., 1982, Self-splicing RNA: autoexcision and autocyclization of the ribosomal RNA intervening sequence of Tetrahymena, Cell 31:147.
2. Takada, C.G., Gardiner, K., Marsh, T., Pace, N., Altman, S., 1983, The RNA moiety of ribonuclease P is the catalytic subunit of the enzyme. Cell 35:849.
3. Forster, A., Symons, R.H., 1987, Self-cleavage of plus and minus RNAs of a virusoid and a structural model for the active site, Cell 49:211.
4. Epstein, L.M., Gall, J., 1987, Transcripts of newt satellite DNA autocleave *in vitro*, Cold Spring Harbor Symp. Quant. Biol. 52:261.
5. Fedor, M., Uhlenbeck, O.C., 1990, Substrate sequence effects on "hammerhead" RNA catalytic efficiency, Proc. Nat. Acad. Sci. USA 87:1668.
6. Uhlenbeck, O.C., 1987, A small catalytic oligoribonucleotide, Nature 328:596.
7. Haseloff, J., Gerlach, W.L., 1988, Simple RNA enzymes with new and highly specific endoribonuclease activity, Nature 334:585.
8. Ruffner, D.E., Stormo, G.D., Uhlenbeck, O.C., 1990, Sequence requirements of the hammerhead RNA self-cleavage reaction, Biochemistry 29:10695.
9. Chang, P.S., Cantin, E.M., Zaia, J.A., Ladne, P.A., Stephens, D.A., Sarver, N., Rossi, J.J., 1990, Ribozyme-mediated site-specific cleavage of the HIV-1 genome, Clinical Biotechnology 2:23.
10. Rossi, J.J., Cantin, E.M., Zaia, J.A., Ladne, P.A., Chen, J., Stephens, D.A., Sarver, N., Chang, P.S., 1990, Ribozymes as therapies for AIDS, AIDS: anti-HIV agents, therapies and vaccines, Annals of the N.Y. Acad. Sci. 616:184.

11. Sarver, N., Cantin, E., Chang, P., Ladne, P., Stephens, D., Zaia, J., Rossi, J.J., 1990, Ribozymes as potential anti-HIV-1 therapeutic agents, Science 247:1222.
12. Saiki, R.K., Scharf, S., Faloona, F., Mullis, K., Horn, G., Erlich, A., Arnheim, N., 1985, Enzymatic amplification of beta-globin genomic sequences and restriction site analysis for diagnosis of sickle cell anemia, Science 230:1350.
13. Leonetti, J., Machy, P., Degols, G., Lebleu, B., and Leserman, L., 1990, Antibody targeted liposomes containing oligodeoxyribonucleotides complementary to viral RNA selectively inhibit viral replication, Proc. Natl. Acad. Sci. USA 87:2448.
14. Chatterjee, S., Wong, K.K., Rose, J.A., Johnson, P.R., 1991, Transduction of intracellular resistance to HIV production by an adeno-associated virus-based antisense vector. IN: Vaccine 91: Modern approaches to new vaccines including the prevention of AIDS. R.M. Channock, H.S. Ginberg, F. Brown and R.A. Lerner, eds. Cold Spring Harbor Laboratory Press, Cold Spring Harbor.
15. Riesner, D., Gross, H.J., 1985, Viroids, Ann. Rev. Biochem. 54:531.
16. Kotin, R.M., Siniscalo, M., Samulshi, J., Zhu, X., Hunter, L., Laughlin, C., McLaughlin, S., Mazycka, N., Rocchi, M., Berns, K.I., 1990, Site-specific integration by adeno-associated virus, PNAS 87:2211.

THERAPIES FOR HEPATITIS B VIRUS: CURRENT STATUS AND FUTURE POSSIBILITIES

Paul Martin and Lawrence S. Friedman

Division of Gastroenterology and Hepatology
Jefferson Medical College
Philadelphia, PA 19107

SUMMARY AND INTRODUCTION

On a world-wide basis the hepatitis B virus (HBV) is the most significant viral pathogen affecting man; several hundred million people are infected. Although the chief burden of illness is felt in certain areas such as the Far East and sub-Saharan Africa, HBV is also a major cause of morbidity and mortality in the United States (1). Estimates by the Centers for Disease Control suggest that there are more than three hundred thousand new cases of acute hepatitis B in the United States every year, with four thousand deaths due to cirrhosis and one thousand deaths due to hepatocellular carcinoma annually (2). Approximately 0.5% of the United States population is chronically infected with HBV, in contrast to 20% of the population in endemic areas such as the Far East and sub-Saharan Africa.

The greater prevalence of HBV infection in Africa and the Far East compared to the United States reflects differing modes of HBV transmission in these areas. In the Far East, perinatal transmission from an HBV carrier mother and, in Africa, perinatal and early childhood transmission from infected family members or playmates is exceedingly common. It is postulated that the immune system of an infant or young child is less effective than that of an adult in mounting a vigorous immune response to HBV, thereby resulting in chronic infection as the most usual outcome of exposure to HBV in this group. In contrast, in Caucasian adults who develop acute hepatitis B, only a minority, about 5%, develop chronic HBV infection.

Chronic HBV infection can result in a number of potential outcomes. These include complete resolution of infection (seroconversion from HBsAg to anti-HBs); cessation of viral replication (seroconversion from HBeAg to anti-HBe) but persistence of hepatitis B surface antigen (HBsAg) (asymptomatic carrier state); or prolonged continuing viral replication manifested serologically by the persistence of hepatitis B e antigen (HBeAg), HBV DNA, and HBV DNA polymerase (3). Prolonged chronic hepatitis B with ongoing viral replication can in turn lead to continuing hepatic damage with histologic evolution from chronic active hepatitis to cirrhosis and potentially to hepatocellular carcinoma (HCC). The development of HCC appears to be particularly common in individuals infected with HBV early in life, in whom the risk of developing HCC is increased two hundred-fold compared to persons without chronic HBV infection (4).

Innovations in Antiviral Development and the Detection of Virus Infection
Edited by T. Block *et al.*, Plenum Press, New York, 1992

Until recently, the presence of HBeAg in the serum was felt to be a reliable marker of viral replication, while the absence of HBeAg was felt to indicate cessation of viral replication. However, with increasing awareness of the molecular sophistication of HBV, it has become apparent that some individuals with chronic hepatitis B may have ongoing HBV replication in the absence of HBeAg (5). Application of the polymerase chain reaction to detect and amplify minute quantities of HBV DNA has indicated that HBV replication can occur at a low level in some patients without standard serological or biochemical markers of viral replication (6). Moreover, there exist mutant forms of HBV that do not produce HBeAg even during active viral replication in the liver (5).

Conventionally, chronic hepatitis has been classified histologically as chronic persistent hepatitis, chronic active hepatitis, and cirrhosis, on the basis of the degree of inflammatory activity and fibrosis. Although these categories are of prognostic value in estimating survival rates in patients with chronic hepatitis, the distinction between the various categories is not always clear cut in an individual case of chronic hepatitis B, in part because of sampling variation on routine percutaneous needle biopsies of the liver and in part because milder forms of chronic hepatitis may evolve to more severe forms on occasion. Thus, patients with generally benign chronic persistent hepatitis due to HBV infection have been reported to progress through the stage of chronic active hepatitis to cirrhosis with liver failure (3). A study from Stanford University has drawn attention to the potentially serious long-term outcome of chronic HBV infection. Patients with chronic active hepatitis and cirrhosis had an estimated 55% five year survival in contrast to those with chronic active hepatitis alone in whom the five-year survival was 86% and those with chronic persistent hepatitis in whom the five-year survival was 97% (7). In addition, the specter of primary hepatocellular carcinoma shadows all patients with chronic HBV infection (4).

RATIONALE AND CURRENT GOALS OF THERAPY

The ideal therapy for chronic HBV infection would result in the complete clearance of HBV from the infected host and the complete resolution of the histological damage. Unfortunately, these goals have been extremely elusive, and therapy at present is directed predominantly towards converting chronic HBV infection with active viral replication to an asymptomatic carrier state without active viral replication. The conventional marker used to assess cessation of HBV replication has been HBeAg in serum. Increasingly, serum HBV DNA levels have emerged as a more precise indicator of intrahepatic HBV replication.

There are a number of unanswered questions about the natural history of chronic HBV infection and optimal therapy. Although loss of serum HBeAg, either spontaneously or with antiviral therapy, has been associated with histological improvement on serial liver biopsies, it is still unclear whether halting HBV replication will ultimately prevent the development of cirrhosis and hepatocellular carcinoma in chronically infected individuals. It is also unclear whether any outcome short of complete loss of HBsAg and development of immunity will prevent future reactivation of HBV replication. Indeed, recent observations suggest that even after HBsAg to anti-HBs seroconversion spontaneously or following anti-viral therapy, HBV DNA may be detected in the liver with the polymerase chain reaction; the clinical significance of these finding remains unclear (8).

After a number of early attempts to treat HBV infection with a variety of antiviral and immunomodulating agents, a role for alpha-interferon in the treatment of HBV has emerged in recent studies (9,10,11). There are now grounds for cautious optimism that at least some patients may benefit from currently available treatment strategies.

APPROACHES TO TREATMENT PRIOR TO ALPHA-INTERFERON

Corticosteroids, which have long been successfully used in the treatment of idiopathic (autoimmune) chronic active hepatitis, have been recognized to be of no value and potentially deleterious in patients with chronic HBV infection when used alone (12,13). Use of antiviral agents such as adenine arabinoside and its more soluble monophosphate derivative (Ara-AMP) initially appeared to be of benefit in chronic HBV infection, but properly controlled clinical trials did not support these initial observations and also drew attention to the considerable toxicity of the drug (14,15). Acyclovir has been suggested as a useful antiviral agent in chronic HBV infection (16), but its antiviral effect against HBV is at best weak and its role in the treatment of chronic hepatitis B is unestablished. Other approaches such as the use of interleukin-2 have also not been shown to be useful in limited experience (17).

THEORETICAL BASIS FOR THE USE OF INTERFERON IN CHRONIC HBV INFECTION

A number of potentially important defects in immune response have been identified in patients with chronic HBV infection. For example, serum levels of interferon are often undetectable despite the presence of continuing viral replication in the liver. Absence of HLA (class 1) antigen expression on hepatocyte membranes, lack of cellular 2',5'-oligoadenylate synthetase production and defective activation of peripheral mononuclear cells are often observed and are consistent with lack of endogenous interferon production. Some patients with chronic HBV infection have been shown to have increased suppressor T cell function and decreased production of immunoglobulins by B lymphocytes in response to exogenous alpha interferon, as compared to normal individuals (18).

With the recognition that immune factors are important in the persistence of chronic HBV infection and the identification of interferon as an important mediator of the host response to viral infections, interferons were thought to be ideal candidates for possible antiviral therapy in chronic HBV infection. There are three major types of interferon: alpha-interferon, produced by B lymphocytes and monocytes; beta-interferon, produced by fibroblasts in response to a virus or polyribonucleotide; and gamma-interferon, produced by T cells in response to a specific antigenic stimulus. Alpha and beta interferon are products of genes located on chromosome 9 in humans, whereas gamma-interferon is produced by a single gene on chromosome 12. As well as sharing a significant degree of homology, the alpha and beta interferons bind to a common receptor which is coded by a gene located on chromosome 21. The gamma-interferon receptor is coded by a gene located on chromosome 6. Both types of interferon receptors are widely distributed on a variety of human cells.

Binding of the various types of interferon to cells results in induction of a number of proteins, including 2',5'-oligoadenylate synthetase, which stimulates an RNAse that cleaves viral RNA preferentially, protein kinase, which also has antiviral effects, and a so-called "MX protein," which in some mammalian species has been shown to confer resistance to viral infection. Some important cell surface proteins are also induced by the binding of interferon to cell receptors. These surface proteins include class I HLA antigens and beta-2 microglobulin. Gamma-interferon has some additional effects on the expression of cell surface proteins, including induction of class II HLA antigens. A variety of other important immune mechanisms are stimulated by interferon activity including regulation of cytokine production and, in the case of gamma-interferon, an increase in macrophage killing and natural killer cell activity (19,20).

Early therapeutic efforts using the various interferons were complicated by the necessity to isolate these proteins from biological systems. With the advent of recombinant DNA technology, it became possible to synthesize interferons in a pure form without contamination by other biologically active products.

Recognition of the immune derangements in chronic HBV infection made the interferons a natural choice for studies of antiviral therapy with alpha-interferon, the most extensively investigated. Beta-interferon and gamma-interferon also appear to have antiviral activity in therapy of chronic hepatitis B. The major problem with all the interferons has been a distressingly high frequency of side effects even in the relatively modest doses used in studies of chronic hepatitis B (see Table 1). Gamma-interferon appears to be particularly associated with these side effects (20).

Table 1. Side Effects of Alpha Interferon

Common	Uncommon
Fatigue	Autoimmune thyroiditis
Flu-like symptoms	Increased propensity to bacterial infection
Anorexia	Overt psychiatric disturbance
Insomnia	
Weight loss	
Altered mood	
Alopecia	
Bone marrow depression	

THERAPEUTIC USE OF INTERFERON IN HBV INFECTION

Although the initial report suggesting a possible therapeutic benefit of interferon in chronic HBV infection appeared in the mid-1970's (21), it has only been within the last few years that controlled clinical trials have demonstrated the efficacy of alpha-interferon in the treatment of chronic HBV infection (9,10,11). With the advent of recombinant DNA technology, allowing production of large quantities of alpha-interferon and the recognition of possible interferon deficiency in chronic HBV infection, clinical trials with alpha-interferon got underway in the 1980's. Preliminary studies have indicated that at least three months' therapy is required for a successful outcome (HBeAg to anti-HBe seroconversion). Generally, side effects were tolerable at dosages of 5 to 10 million units daily by subcutaneous injection, and in the initial studies these dosages appeared to be as efficacious as larger doses (19, 22).

Initial controlled studies indicated a successful response with loss of HBeAg and improvement in liver histology in approximately one-third of patients.[9] Subsequently, Perrillo reported that the addition of prednisone "priming" for six weeks prior to treatment with alpha-interferon boosted the response rate to approximately 50%. The rationale for pretreatment with prednisone was that it enhanced viral replication and immune-responsiveness, leading to a better response to subsequent alpha-interferon therapy[10]. It was recognized that the use of prednisone in patients with chronic hepatitis B was possibly hazardous given previous reports documenting severe hepatic decompensation when prednisone was administered to HBV-infected individuals with advanced chronic liver disease [12, 13].

A subsequent larger study, also reported by Perrillo (11), has clarified and amplified these finding. It appears that priming prednisone therapy is only a benefit to patients with low pre-treatment serum alanine aminotransferase activities (<100 U/L) (11). Prednisone pretreatment does not appear to enhance the response rate to alpha-interferon in individuals who already have a brisk immune-response as manifested by elevated serum aminotransferase levels and marked inflammatory activity on liver biopsy. This report also drew attention to the possible risks of combination therapy, in that one patient with cirrhosis succumbed from hepatic decompensation due to reactivation of HBV infection receiving prednisone. In fact, patients who have a successful response to interferon (even without prednisone priming) have a rise in serum aminotransferase levels several weeks into their course of interferon, presumably due to increased immune activation. This phenomenon is another potential trigger for hepatic decompensation (12,13).

Thus, the status of therapy for chronic HBV infection at present is that alpha-interferon benefits just under half the patients treated alone or, in selected cases, in combination with prednisone (1). In Perillo's studies (10,11), approximately one-third of patients who lost HBeAg also ultimately lost HBsAg, suggesting complete viral clearance. This implies that overall, only a minority of treated patients will have complete resolution of HBV infection. Moreover, as noted above, even persons who undergo HBsAg to anti-HBs seroconversion following therapy with interferon or spontaneously, may have persistent intrahepatic HBV DNA detectable by polymerase chain reaction, a finding of uncertain significance. However, these patients had no evidence of transcriptionally active HBV as shown by the absence of HBV-related messenger RNA. In addition, the persistence of hepatic HBV DNA did not appear to preclude histological improvement after loss of HBsAg (8).

Although these results are considerably better than those for any previously reported form of therapy, the limitations of a treatment regimen that produces a sustained response in less than half the treated patients are obvious (see Table 2). In addition, although alpha-interferon in the dosages used was reasonably well tolerated, it appears that higher doses are associated with increased toxicity and do not markedly increase the chance of response (23). Combining interferon with antiviral agents such as acyclovir and Ara-AMP has been reported but does not appear to enhance the overall therapeutic response. Thus it appears that alpha-interferon, although promising, is by no means a panacea in the treatment of chronic HBV infection and future efforts will likely be directed toward alternative, novel forms of antiviral therapy.

Another unanswered question is whether the reported results with interferon can be reproduced in other groups with a high prevalence of chronic HBV infection, including Orientals infected at an early age (24) and individuals who are infected simultaneously with HBV and the human immunodeficiency virus (HIV) (25).

Table 2. Prediction of Poor Outcome in HBV Therapy

1. Prolonged duration of infection especially acquisition at an early age.
2. Absence of vigorous immune response (high HBV DNA, no or modest aminotransferase elevation, absence of marked hepatic inflammation).
3. Male sex
4. HIV positive

Recently, encouraging results have been reported in the treatment of Orientals with chronic HBV infection with interferon, and it is hoped that with better patient selection a response comparable to that observed in Caucasians with chronic HBV will be observed (24). However, as demonstrated in Perrillo's most recently reported study (11), an important determinant of responsiveness to interferon is the duration of HBV infection, and because Orientals in endemic areas are predominantly infected with HBV as infants, responsiveness of adult Orientals with chronic hepatitis B to interferon may be less than that seen in Caucasians infected as adults. Differences in responsiveness to interferon may be explained at least in part by the lack of integration of HBV into the host genome early in chronic HBV infection; this would favor a response to interferon in adults who acquire HBV infection as adults rather than as infants or children.

There are also suggestions in the literature that male homosexuals with chronic HBV infection are less likely to respond to alpha-interferon treatment than heterosexual males (25,26). A major confounding factor in these comparative observations is a higher rate of HIV infection in male homosexuals but even when HIV status is taken into account, homosexuals appear to have a lower response rate to interferon than heterosexuals (11). In general, females with chronic hepatitis B seem to have a better chance of responding to alpha-interferon that males, and it is not clear why the sexual orientation of males uninfected with HIV should be a determinant of response.

Alpha-interferon therapy of chronic HBV infection is generally ineffective in patients who are immunosuppressed, either therapeutically or pathologically (as in AIDS) (27). Interferon has not been proven to be of benefit in preventing HBV reinfection of or in treating recurrent hepatitis B in liver allograft recipients. In addition, there is concern that by encouraging expression of class I HLA antigens on hepatocytes, interferon may promote graft rejection.

ANIMAL MODELS OF HBV INFECTION

One area of great interest in the development of new therapies for, as well as understanding the biology of, HBV is the recognition of viruses similar to HBV which infect nonhuman species. These hepadnaviruses include viruses that affect other mammals; the woodchuck hepatitis virus and the common ground squirrel hepatitis virus, as well as at least two avian-infecting species whose hosts are the Peking duck and Grey heron (28).

Several of these animal species have been studied extensively as models of hepadnavirus virology, associated liver disease, and hepatocellular carcinoma. More recently, a number of investigators have reported on the use of the duck hepatitis B model to develop new antiviral agents (29).

A seminal observation has been that hepadnaviruses replicate in a manner similar to retroviruses (2). Specifically, hepadnaviruses, like retroviruses, employ a reverse transcriptase. This recognition has coincided with a concerted effort to find an effective antiviral agent for the treatment of patients with HIV. Among a variety of agents studied, nucleoside analogues, including 3'azido-2'3'-deoxythymidine (AZT) have already gained wide acceptance in the treatment of HIV infection (31). One nucleoside analogue that has also been studied extensively in the duck hepadnavirus model is dideoxycytidine (ddC). This agent was selected for study because of its similarity to the prototype of this group of compounds, AZT, with the hope that it would be less toxic than AZT in human subjects. Unfortunately, ddC appears to have considerable toxicity of its own. However, in vivo testing in the duck model demonstrated considerable efficacy as manifested by inhibition of duck viral replication. A short period of intravenous ddC therapy resulted in a fall in markers of viral replication including serum DNA polymerase and duck virus DNA. These initial observations have subsequently been expanded using other members of the dideoxynucleoside family (32), but whether these agents will be applicable to the treatment of chronic hepatitis B in humans remains uncertain.

ROLE OF COMBINATION THERAPY IN HBV INFECTION

A number of investigators have tried using combinations of agents to increase the efficacy of currently available antiviral therapy. To date, the only effective combination appears to be prednisone "priming" followed by alpha-interferon in patients with only modest serum aminotransferase elevations. Other combinations tried in the past but discarded have included prednisone with Ara-AMP as well as Ara-AMP with interferon. There are currently trials underway combining acyclovir with interferon (33).

If the initial promise of the dideoxynucleosides is fulfilled, it is possible that there may be a role for these agents in combination with interferon. By halting viral replication, at least in the duck hepadnavirus model, the dideoxynucleosides decrease the amount of circulating viral DNA. Therefore, in combination with the immunomodulating and direct antiviral effects of interferon, the dideoxynucleosides may possibly result in an enhanced response to treatment. Proof of this awaits further clinical trials.

The direct viral inhibitory effects of the dideoxynucleosides may be particularly useful in liver allograft recipients. Reinfection of allografts with HBV has been one of the most difficult issues in hepatic transplantation. Alpha-interferon appears to lack efficacy and may even enhance the risk of graft rejection because of enhanced expression of class I HLA antigen on cell surfaces (27). Direct inhibition of viral replication by agents such as dideoxynucleosides could conceivably prevent reinfection of the graft by preventing viral replication in extrahepatic sites for prolonged periods of time following successful liver transplantation.

SUMMARY

Alpha-interferon is the first agent with proven therapeutic efficacy in humans with chronic HBV infection. Obviously, further research is needed to clarify and expand the role of interferon in this setting. In addition, an overall response rate of less than 50% illustrates the need for continuing innovation in the treatment of HBV infection especially in patients with predictors of poor outcome. Despite the advent of effective vaccines against HBV, this pathogen is likely to remain a source of serious human morbidity and mortality for the foreseeable future. Continuing efforts must be directed towards finding more effective therapies against HBV. A recent preliminary report suggests that thymosin. like interferon on immune modulator, may also be efficacious in the treatment of chronic HBV and deserves further study[34, 35].

REFERENCES

1. Hoofnagle, J.H., Chronic hepatitis B, N. Engl. J. Med. 323:337-339 (1990).
2. Aach, R.D., The treatment of chronic type B viral hepatitis, Ann. Intern. Med. 109:89-91 (1988).
3. Seeff, L.B., Koff, R.S., Evolving concepts of the clinical and serological consequences of hepatitis B virus infection, Sem. Liv. Dis. 6:11-22 (1986).
4. Beasley, R.P., Hwang, E.Y., Lin, C-C., Chien, C-S., Hepatocellular carcinoma and hepatitis B virus, Lancet ii:1129-1133 (1981).
5. Carman, W.F., Jacyn, M.R., Hadziyannis, S., Karayiannis P., McGarney M.J., Makris A., Thomas H.C., Mutation preventing formation of hepatitis B e antigen in patients with chronic hepatitis B infection, Lancet ii: 88-591 (1989).
6. Kaneko, S., Miller, R.H., Feinstein, S.M., Unura M., Kobayashi K., Hatlor N., Detection of serum hepatitis B virus DNA in patients with chronic hepatitis using the polymerase chain assay, Proc. Natl. Acad. Sci. USA 86:312-316 (1989).
7. Weissberg, J.I., Andres, L.J., Smith, C.I., Weick S., Nichols J.E., Garcia G., Robinson W., Merigan T., Gregory P., Survival in chronic hepatitis B: an analysis of 379 patients, Ann. Intern. Med. 101:613-616 (1984).
8. Kuhns, M., Mason, A., McNamara, A., Campbell, C., Perrillo R.P., Polymerase chain reaction in the determination of HBV DNA after HBsAg clearance in chronic hepatitis B, Hepatology 12:904 (1990).
9. Hoofnagle, J.H., Peters, M., Mullen, K.D., Jones D.B., Rustyi V., Di Bisceglie A., Itallahan C., Park Y., Meschievitz C., Jones E.A., Randomized controlled trial of recombinant human alpha interferon in patients with chronic hepatitis B, Gastroenterology 95:1318-1325 (1988).
10. Perillo, R.P., Regenstein, F.G., Peters, M.G., Deschryver-Kecskemeti K., Bodicky C., Campbell C.R., Kuhns M.C., Prednisone withdrawal followed by recombinant alpha interferon in the treatment of chronic type B hepatitis: a randomized, controlled trial, Ann. Intern. Med. 109:95-100 (1988).
11. Perillo, R.P., Schiff, E.R., Davis, G.L., Bodenheimer H.C. Jr., Lindsay K., Pagne J., Dienstag J.C., O'brien C., Tamburro C., Jacobson I.N., Sampliner R., Feit D., Lefkowitch J., Kohns M., Meschievitz C., Sanghri B., Albrecht J., Gibas A., A randomized controlled trial of interferon alpha-2b alone and after prednisone withdrawal for the treatment of chronic hepatitis B, N. Engl. J. Med. 323:295-301 (1990).
12. Lam, K.L., Lai, C.L., Trepo, C., Wu, P.C., Deleterious effect of prednisone in HBsAg-positive chronic acute hepatitis, N. Engl. J. Med. 304:380-386 (1981).
13. Hoofnagle, J.H., Davis, G.L., Pappas, S.C., Hanson R., Peters M. Avigan M., Waggoner J., Jones E.A., Seeff L.B., A short course of prednisone in chronic type B hepatitis: report of a randomized, double-blind, placebo-controlled trial, Ann. Intern. Med. 104:12-17 (1986).

14. Garcia, G., Smith C.C., Weissberg, J.I., Eisenberg M., Bissett J., Nair P.V., Mastre B., Rosno S., Roskamp D., Waterman K., Pollard R.B., Tong M.J., Brown B.W., Robinson W.S., Gregory P.B., Merigan T.C., Adenine arabinoside monophosphate (vidarabine phosphate) in combination with human leukocyte interferon: a randomized, double-blind, placebo-controlled trial, Ann. Intern. Med. 107:278-285 (1987).
15. Hoofnagle, J.H., Hanson, R.G., Minuk, G.Y., Pappas S.C., Schafer D.F., Dusheiko G.M., Straus S.E., Popper H., Jones E.A., Randomized controlled trial of adenine arabinoside monophosphate for chronic type B hepatitis, Gastroenterology 86:150-157 (1984).
16. Alexander, G.J.M., Fagan, E.A., Hagarty, J.E., Yeo J., Eddleston A.C., Williams R., Controlled clinical trial of acyclovir in chronic hepatitis B virus infection, J. Med. Virol. 21:81-87 (1987).
17. Kakumu, S., Fuju, A., Yoshioka, K., Tahara H., Ohtani Y., Hirofuji H., Murase K., Aoi, T., Pilot study of recombinant human interleukin-2 for chronic type B hepatitis, Hepatology 8:487-492 (1988).
18. Peters, M, Immunological aspects of antiviral therapy, Springer Sem. Immunopathol. 12:47-56 (1990).
19. Davis, G.L., Hoofnagle, J.H., Interferon and viral hepatitis, Hepatology 5:1038-1041(1986).
20. Di Bisceglie, A.M., Rustgi, V.K., Kassianides, C., Lisker-Melman M., Park Y., Waggoner J.G., Hoofnagle J., Therapy of chronic hepatitis B with recombinant human alpha and gamma interferon, Hepatology 11:266-270 (1990).
21. Greenberg, H.B., Pollard, R.B., Lutwick, L.I., Gregory P.B., Robinson W.S., Merigan, T.C., Effect of human leukocyte interferon on hepatitis B virus infection in patients with chronic active hepatitis, N. Engl. J. Med. 295:517-522 (1976).
22. Perrillo, R.P., Interferon therapy of hepatitis B infection, Sem. Liv. Dis. 9:240-248 (1989).
23. Scully, L.J., Shein, R., Korayiannis, P., McDonald J.A., Thomas H.C., Lymphoblastoid interferon therapy of chronic HBV infection: a comparison of 12 vs. 24 weeks of thrice weekly treatment, J. Hepatol. 5:51-58 (1987).
24. Lok, A.S.F., Lai, C.L., Wu, P.C., Lan J.Y., Leung E.K., Wong L.S., Treatment of chronic hepatitis B with interferon: experience in Asian patients, Sem. Liv. Dis. 9:249-253 (1989).
25. Novick, D.M., Lok, A.S.F., Thomas, H.C., Diminished responsiveness of homosexual men to antiviral therapy for HBsAg-positive chronic liver disease, J. Hepatol. 1:29-35 (1984).
26. McDonald, J.A., Caruso, L., Korayiannis, P., Scully L.J., Harris J.R.W., Forster G.E., Thomas H.C., Diminished responsiveness of male homosexual chronic hepatitis B virus carriers with HTLV-III antibodies to recombinant alpha-interferon, Hepatology 7:719-723 (1987).
27. Davis, G.L., Interferon treatment of viral hepatitis in immunocompromised patients, Sem. Liv. Dis. 9:267-272 (1989).
28. Mason, W.S., Taylor, J.M., Experimental systems for the study of hepadnavirus and hepatitis delta virus infection, Hepatology 9:635-645 (1989).
29. Kassianides, C., Hoofnagle, J.H., Miller, R.H., Doo E., Fordi H., Broder S., Mitsuya H., Inhibition of duck hepatitis B virus replication by 2',3'-dideoxycytidine: a potent inhibitor of reverse transcriptase, Gastroenterology 97:1275-1280 (1989).
30. Summers, J., Mason, W.S., Replication of the genome of a hepatitis B-like virus by reverse transcriptase of an RNA intermediate, Cell 29:403-415 (1982).
31. Yarchoan, R., Mitsuya, H., Meyers, C.E., Broder, S., Clinical pharmacology of 3'-azido-2'3'-deoxythymidine (Zidovidine) and related dideoxynucleosides, N. Engl. J. Med. 321:726-738 (1989).

32. Martin, P., Kassianides, C., Korenman, J.E., Hoofnagle J.H., Ford H., Broder S., Mitsuya H., 2'-3'-dideoxyinosine and dideoxyguanosine are potent inhibitors of hepadnaviruses in vivo, Gastroenterology 96:628 (1989).
33. Schalm, S.W., Heyfink, R.A., Van Burnes, H.R., De Man, R.A., Acyclovir enhances the antiviral effect of interferon in chronic hepatitis B, Lancet ii:358-360 (1985).
34. Mutchick, M.G., Appelman, H.D., Chung, H.T., Aragone, E., Grupta, T.P., Cummings, G.D., Waggoner, J.G., Hoofnagle, J.M., Shiefritz, D.A. Thymosin treatment of chronic hepatitis B: A placebo-controlled pilot trial. Hepatology 14:409-415 (1991).
35. Davis, G.L. Treatment of chronic hepatitis B, Hepatology 14:567-569 (1991).

A PRELIMINARY REPORT OF A CONTROLLED STUDY OF THYMOSIN ALPHA-1 IN THE WOODCHUCK MODEL OF HEPADNAVIRUS INFECTION

John L. Gerin*, Brent E. Korba*,
Paul J. Cote*, and Bud C. Tennant+

*Division of Molecular Virology and Immunology
Department of Microbiology
Georgetown University
Rockville, Maryland

+Department of Clinical Sciences
College of Veterinary Medicine
Cornell University
Ithaca, New York

INTRODUCTION

Thymosin alpha-1 (TA1) is one of a family of peptides with immunoregulatory activities which occur naturally in thymic tissues (1). TA1 is an acidic 28 amino acid molecule that is biologically active in a number of *in vitro* and *in vivo* assays; it is highly conserved among mammalian species and is available as a chemically-defined synthetic preparation. Detectable levels of TA1 are found in the serum of normal individuals (2). Due to the immunoregulatory properties of thymic peptides, there is current interest in their possible use in the treatment of chronic diseases, including Type B hepatitis (3). In a preliminary report, Mutchnick, et al. (4) described the possible efficacy of thymosin fraction 5 and TA1 in the treatment of patients with chronic active hepatitis due to hepatitis B virus (HBV). The purpose of the current study was to evaluate TA1 in an animal model system of HBV infection and disease.

The eastern woodchuck (*Marmota monax*) is naturally infected with the woodchuck hepatitis virus (WHV), a member of the hepadnavirus family and closely related to HBV. The woodchuck has been developed as an experimental model of chronic WHV infection; chronically-infected animals progress to chronic liver disease, including hepatocellular carcinoma (HCC) in a predictable manner (5). Knowledge of the natural history of WHV infection in the experimental model permits its use in the evaluation of potential antiviral and immunomodulatory strategies. As TA1 is highly conserved among mammalian species, this model is an appropriate one in which to complement the above-mentioned trial of TA1 in man.

Innovations in Antiviral Development and the Detection of Virus Infection
Edited by T. Block *et al.*, Plenum Press, New York, 1992

STUDY DESIGN

In preliminary studies with normal woodchucks, a dose of 10 ug/Kg given subcutaneously produced a peak TA1 blood concentration of approximately 100 ng/ml within an hour of administration with a decline to basal levels (1-2 ng/ml) within 6-12 hours. This dose was judged to be suitable for an antiviral trial of TA1. Twelve experimentally-derived chronic WHV carrier woodchucks were used in the study; all animals had histologic evidence of chronic hepatitis in the pretreatment biopsy and no evidence of HCC. TA1 was administered subcutaneously (10 ug/Kg body weight) twice-weekly for 30 weeks to six animals. Six age- and sex-matched chronic WHV carrier animals were used as untreated controls. Blood samples were taken from treated animals prior to treatment and at regular intervals during the treatment and six-month follow-up period. Serum was analyzed for markers of WHV infection: WHV DNA, WHsAg, anti-WHc and anti-WHs. Liver biopsies from all animals were taken prior to treatment, during the course of treatment and during the follow-up period; liver tissues were examined histologically and for markers of WHV replication using hybridization-based methods to quantitate WHV DNA and RNA forms. Standard clinical parameters were regularly measured in order to evaluate possible drug toxicity.

RESULTS AND DISCUSSION

During the treatment and follow-up periods, untreated control woodchucks maintained stable levels of serum and liver markers of active WHV replication (serum WHV DNA and hepatocyte WHV replicative intermediates). In contrast, all TA1 treated animals demonstrated a steady decline in levels of viremia during the treatment phase which paralleled the loss or replicative WHV forms in liver. By the end of treatment, 4 of the 6 treated animals had shown a 1000-fold or greater decrease in serum WHV DNA concentration and a 10-fold or greater drop in liver replicative forms. In the remaining two animals, the suppression of WHV markers was less (30 to 90-fold decrease in viremia with corresponding suppression of liver forms). The difference between the control and treated animals in response to TA1 treatment was significant ($p<0.05$).

In the follow-up period, serum and tissue markers of WHV replication in the treated group returned to pretreatment levels in all surviving woodchucks. Three animals in each group died in the follow-up period, five of HCC (3 controls, 2 treated) and 1 treated animal of a ruptured aortic aneurism. There was no difference between the treated and control groups in the degree of chronic hepatitis or development of HCC as an outcome. The time to appearance of HCC was consistent with the natural course of disease from prior studies. No evidence of drug toxicity was observed in the course of the trial.

In summary, a controlled trial in chronic carrier woodchucks demonstrated that a six-month course of TA1 treatment caused a significant suppression of serum and tissue markers of hepadnavirus replication. The effect, however, was transient and viral replication returned to pretreatment levels in the follow-up period. TA1 was well-tolerated and no drug-associated toxicity was observed. TA1 treatment had no detectable effect on the underlying virus-induced chronic hepatitis or development of HCC.

The mechanism of the antiviral effect is unknown but may be associated with the immunoregulatory activity of TA1, as demonstrated in other systems. These results are promising and in some respects confirm the experience in the human trial. Further studies in this model are needed to define the optimum dose and treatment period to accomplish long term and, possibly, permanent suppression of viral replication as a prelude to controlled efficacy trials of TA1 against hepadnavirus-induced disease.

ACKNOWLEDGEMENTS

This work was supported by the NIAID under NO1-AI-72623 and NO1-AI-52585 with Georgetown University and Cornell University, respectively, and as part of the Hepatitis B Antiviral Testing Program of the NIAID. The authors gratefully acknowledge the assistance of Milton Mutchnick, M.D. for serum assays of TA1 and Paul Chretien, M.D. for advice on dose studies. The synthetic thymosin alpha-1 used in these studies was provided by Alpha-1 Biomedicals, Washington D.C. through an agreement with NIAID.

REFERENCES

1. Goldstein, A.L., Low, T.L.K., McAdoo, M., et al., Thymosin alpha 1: isolation and sequence analysis of an immunologically active thymic polypeptide, Proc. Natl. Acad. Sci. USA 74:725-729 (1977).
2. McClure, J.E., Lameris, N., Wara, D.W., and Goldstein, A.L., Immunochemical studies on thymosin: radioimmunoassay of thymosin alpha-1, J. Immunol. 128:368-375 (1982).
3. Eichberg, J.W., Seeff, L.B., Lawlor, D.L., Buskell-Bales, Z., Ishak, K., Hoofnagle, J.H., Goldstein, A.L., and Langloss, J.M., Effect of thymosin immunostimulation with and without corticosteroid immunosuppression on chimpanzee hepatitis B carriers, J. Med. Virol. 21:25-37 (1987).
4. Mutchnick, M.G., Lee, H.H., Haynes, G.D., Hoffnagle, J.H., and Appelman, H.O., Thymosin treatment of chronic active hepatitis B (CAHB): a preliminary report on a controlled double blind study, Hepatology 8:1270 (1988).
5. Gerin, J.L., Cote, P.J., Korba, B.E., and Tennant, B.C., Hepadnavirus-induced liver cancer in woodchucks, Cancer Detection and Prevention 14:227-229 (1989).

DELTA VIRUS AS A VECTOR FOR THE DELIVERY OF BIOLOGICALLY-ACTIVE RNAs: POSSIBLY A RIBOZYME SPECIFIC FOR CHRONIC HEPATITIS B VIRUS INFECTION

Sen-Yung Hsieh and John Taylor

Fox Chase Cancer Center
Philadelphia, PA 19111

SUMMARY AND INTRODUCTION

RNA species have been designed that demonstrate specific biological activities. Examples include the antisense RNA oligonucleotides (Weintraub et al., 1985) and catalytic RNAs that act as endoribonucleases (Haseloff and Gerlach, 1988). And, recently Ellington and Szostak (1990) have developed an even more general method of allowing the selection, in the test tube, of RNA species that can bind to a chosen domain, for example on an RNA, DNA, or protein.

Add to this, that there are now mechanisms for specifically and efficiently delivering such biologically active RNAs. Naked nucleic acids, without or with stability-conferring modifications, can be used. But it is also possible to use liposomes, with or without some kind of targeting specificity. And at yet another level, the RNAs can be introduced via viral systems, such as retroviruses, Sindbis virus or vaccinia virus.

The potential to apply these concepts has attracted those who might wish to specifically knock out the expression of a particular cellular or viral gene. Along such lines, the purpose of our studies has been to determine the feasibility of using hepatitis delta virus (HDV) as a means of delivering biologically active RNAs, with the specific intention of finding out whether an HDV-based vector could be used as part of a therapy for humans with chronic active hepatitis B. For such patients there is currently no successful therapy and there is also a very high associated risk of liver cirrhosis and cancer.

In this report we present a feasibility study in which we have used the HDV genome to deliver an RNA that is biologically active. Also we describe how we might extend this to the therapy of chronic HBV infections.

REPLICATION OF HEPATITIS DELTA VIRUS

In order to understand our study it is necessary to begin with some explanation of the structure and replication of HDV. This agent is actually subviral, in that in its life cycle, it is dependent upon the presence of hepatitis B virus (HBV), to provide an envelope. In all other aspects of HDV replication it would appear that there is no need for HBV sequences or functions (Kuo et al., 1989). The HDV genome is a small single-stranded RNA of about 1700 nucleotides in length. This RNA is circular in conformation, and has the ability to fold on itself by intramolecular base-pairing to form an unbranched rod-like structure (Chen et al., 1986). This genome is synthesized and in turn becomes the template for the synthesis of more genomic RNA (Taylor, 1990a, 1990b).

We have previously shown that HDV RNA replication can occur after DNA transfection using an expression vector with, as insert, a multimer of HDV cDNA. Such replication is just as efficient as during a natural infection and leads to about 300,000 copies of genomic RNA per cell (Chen et al., 1986). With this in mind, we asked whether it might be possible to insert onto the HDV RNA genome some extra sequences, with selected and specific biological activity.

FEASIBILITY STUDY USING A RIBOZYME DIRECTED AGAINST A SPECIFIC mRNA

Our feasibility study consisted of five steps: (1) selection of a site of insertion of a biologically-active sequence on the HDV genome; (2) deciding the nature of this insert; (3) demonstrating that the insertion was not inhibitory to genome replication; (4) determining if the insert remained stably as part of the HDV genome; and (5) proving that the insert was still biologically active when localized on the HDV genome. As explained below we were able to satisfy each of these steps.

First we needed to select a suitable insertion site. In this respect we had to avoid on the genome certain regions that we expected to be inappropriate for insetion. We avoided what we refer to as the top of the rod because it seems to include the promotor for RNA-directed RNA synthesis. Also, the central region encoding the delta antigen, the only known protein of HDV, was known to be essential and was thus avoided. Beyond this were regions we have proven to be necessary for polyadenylation and for self-cleavage (Taylor, 1990a, 1990b). In summary, we were left with only a site near the bottom end of the rod structure.

Second, we had to choose the insert. We designed a small 32 nucleotide sequence based on one that Haseloff and Gerlach (1988) had shown to act *in vitro* as a ribozyme and specifically cleave the mRNA for chloramphenicol acetyl transferase, CAT. We used a PCR procedure to remove 3 nucleotides from the cDNA copy of the HDV genome at position 797, using the nomenclature of Kuo et al. (1988), and insert the 32 nucleotide ribozyme, See fig. 1.

Third, we assembled a head-to-tail dimer of this modified HDV cDNA into a eukaryotic expression vector, pSVL (Kuo et al., 1989). This recombinant was transfected into COS7 cells and tested for its ability to initiate RNA-directed HDV RNA synthesis. We found that it replicated just as well as the normal, unmodified, HDV cDNA, as previously reported by Kuo et al.

The fourth question was to test whether the modified HDV genome was able to maintain its insert during genome replication. This was readily demonstrated using a PCR procedure.

Finally, it was necessary to show that the RNA insert when present on the HDV RNA genome was able to function as a ribozyme and cleave the CAT mRNA. To do this we first had to cotransfect cells with another plasmid, that would drive the expression of CAT mRNA. We used two different vectors to express CAT. They differed only in the strength of the associated promotor for pol II-directed synthesis of CAT mRNA; one was driven by the strong immediate-early promotor of cytomegalovirus, CMV-IE, and the other by the relatively weak thymidine kinase promotor of Herpes simplex virus, HSV-TK. Two assays were used to monitor the effect on the CAT mRNA. The first was an S1 nuclease assay and the other was an enzymatic assay for the acetylation of chloramphenicol. With these assays we demonstrated that the presence of the modified HDV RNA genome was able to achieve a three-fold suppression of CAT expression.

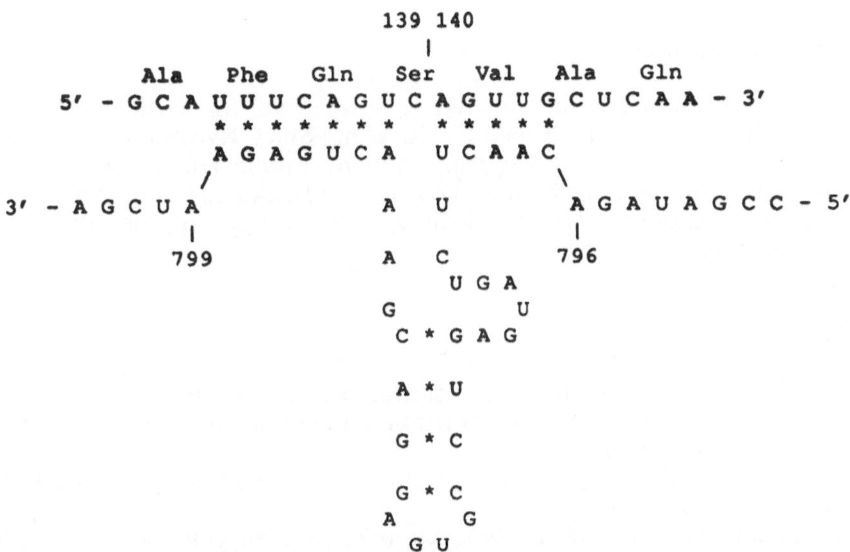

Figure 1. A ribozyme inserted into the genomic RNA of HDV, and directed against the mRNA for chloramphenicol acetyl transferase. The upper line shows that portion of the coding region of the CAT mRNA at which the ribozyme should act to bring about self-cleavage (site indicated by vertical line). The ribozyme, based on that designed by Haseloff and Gerlach (1988), is inserted between nucleotides 796 and 797 on the genomic RNA or HDV, using the number of Kuo et al. (1988).

EVALUATION

We consider the above results as promising. Perhaps the insert could be modified so as to achieve even greater levels of inhibition. Also, we need to confirm that the inhibory action was actually via ribozyme action; after all, as can be seen in Figure 1, the ribozyme contains a short stretch of CAT antisense sequence that could provide some, or all, of the observed effect.

POTENTIAL CLINICAL APPLICATION

We are already moving ahead with a different insertion. This one is directed against a part of the RNA pregenome of HBV. We have deliberately opted for a site on HBV that is shared with the related virus of woodchucks, woodchuck hepatitis virus, WHV. This will make it possible for us to test the action *in vitro*, using transfection of cell lines. Also it will allow us to test the action on woodchucks chronically infected with WHV prior to moving to HBV infections of primates. One important advantage of HDV for the animal studies is delivery. Preliminary experiments indicate that we can package HDV RNA genomes into virus particles using cell cultures (Wang-Shick Ryu and J.M.T.). Such particles, containing genomes with the anti-HBV specific insert, could thus be readily delivered to an animal and would correctly target to the HBV-infected heatocytes.

ACKNOWLEDGEMENTS

We thank William Mason and Wang-Shick Ryu for critical reading of the manuscript, and John Burch, Dorene Davis, Wang-Shick Ryu and Christopher Seeger for experimental advice. J.T. was supported by Public Health Service grants AI-26522, CA-06927, and RR-05539 from the National Institutes for Health, by grant MV-7M from the American Cancer Society, and by an appropriation from the Commonwealth of Pennsylvania.

REFERENCES

Chen, P.J., Kalpana, G., Goldberg, J., Mason, W., Werner, B., Gerin, J., and Taylor, J., 1986, Structure and replication of the genome of hepatitis delta virus, Proc. Natl. Acad. Sci. USA, 83:8774.

Ellington and Szostak, 1990, *In vitro* selection of RNA molecules that bind specific ligands, Nature, 346:818.

Haseloff, J., and Gerlach, W.L., 1988, Simple RNA enzymes with new and highly specific endoribonuclease activities, Nature, 334:585.

Kuo, M.Y.-P., Chao, M., and Taylor, J., 1989, Initiation of replication of the human hepatitis delta virus genome from cloned DNA: Role of the delta antigen, J. Virol., 63:1945.

Kuo, M.Y.-P., Goldberg, J., Coates, L., Mason, W., Gerin, J., and Taylor, J., 1988, Molecular cloning of hepatitis delta virus RNA from an infected woodchuck liver: sequence, structure, and application, J. Virol., 62:1855.

Taylor, J., 1990a, The structure and replication of human hepatitis delta virus, Seminars in Virol., 1:135.

Taylor, J., 1990b, Hepatitis delta virus: cis and trans function needed for replication, Cell, 61:371.

Weintraub, H., Izant, J.G., and Harland, R., 1985, Antisense RNA as a molecular tool for genetic analysis, Trends in Gen., 1:22.

INHIBITION OF HERPES SIMPLEX VIRUS RIBONUCLEOTIDE REDUCTASE BY SYNTHETIC NONAPEPTIDES: A POTENTIAL ANTIVIRAL THERAPY

Michel Liuzzi*, Erika Scouten and Rolf Ingemarson

Department of Biochemistry
Bio-Mega Inc.
2100, rue Cunard, Laval, Québec, Canada, H7S 2G5

INTRODUCTION

Herpes simplex virus (HSV) types 1 and 2 are pathogenic DNA viruses that cause diverse diseases in humans[1], perhaps best exemplified by genital herpes which is primarily due to HSV-2 infection. It has been estimated that more than 20 million people are affected with genital herpes in the United States alone and that about 300,000 new cases are reported each year. Since herpesvirus infections are eventually attenuated by the host immune system in otherwise normal persons, newborns and immunocompromised patients are particularly vulnerable to more severe forms of infections. HSV belongs to the family of human herpesviruses which also includes varicella zoster virus (VZV), cytomegalovirus (CMV), Epstein-Barr virus (EBV) and human herpesviruses 6 and 7 (HHV-6 and HHV-7). While these different viruses are responsible for a variety of different pathologies associated with their specific life cycles in different human tissues, they do have several key features in common. After primary infection, herpesviruses have the capability to establish latency[2], which is defined as a state in which the viral genome is present in certain cells but infectious virus is not produced. These viruses can be reactivated from the latent state by various stimuli such as ultraviolet light, stress and fever to produce new infectious virus. With respect to herpesvirus infections, development of vaccines has, thus far, met with little success. Therefore, it is important to develop alternative antiherpetic agents for use in the treatment of herpesvirus-induced diseases.

In order to help develop effective antiviral strategies, it is desirable to fully understand the virus and its reproductive cycle. Recent knowledge at the molecular level of how herpesviruses replicate has made it possible to delineate processes that are unique to the virus and to identify new opportunities for therapy[3]. HSV is believed to bind to an as yet unidentified cellular receptor, after which the viral envelope fuses with the plasma membrane, and the nucleocapsid is liberated into the cytoplasm of the host cell. The viral capsid is degraded and the HSV genome, coding for at least 70 identified genes[4], is released into the nucleus where it is transcribed by cellular RNA polymerase II. Viral genes are expressed in a sequential, coordinated fashion[5]. Regulatory proteins of the immediate-early or α class are produced first, followed by early or β proteins, the latter of which peak at approximately 5-7 hours post-infection. This group of β proteins includes the viral enzymes for nucleotide metabolism and DNA processing such as DNA polymerase, thymidine kinase and ribonucleotide reductase. Late or γ proteins which are synthesized at approximately 15-

18 hours after infection comprise mainly viral structural proteins that are required for assembly of new virus particles.

The most investigated target for chemotherapeutic intervention in the HSV replication cycle has been the viral DNA polymerase. Nucleoside analogues have been of particular interest[6]. In general, these compounds need to be activated by viral and/or cellular enzymes to be converted into DNA polymerase inhibitors. In the case of acyclovir, this synthetic nucleoside must be phosphorylated by the HSV thymidine kinase in order for it to be utilized by the DNA polymerase as a chain terminator, thus interfering with viral DNA synthesis and preventing the production of infectious virus. Recently, however, nucleoside analogues have been facing three major problems which may limit their effectiveness as antiherpetic agents. First, because of functional and/or structural similarities between the viral enzyme and its cellular counterpart many nucleoside analogues lack selectivity and cause adverse effects to host cells. Second, in most cases, these compounds display a limited range of action against the various viruses of the herpes family. This is particularly evident for acyclovir which is not effective against the human cytomegalovirus[7] and for bromovinyldeoxyuridine which is very effective against HSV-1 but not HSV-2[8]. Finally, drug-resistant herpesvirus mutants have been obtained recently from patients treated with nucleoside analogues[9]. For example, the herpes simplex virus can acquire resistance to acyclovir through mutations in its DNA polymerase or thymidine kinase gene. However, it is expected that drug combination regimens may prove valuable in circumventing this latter problem. Nevertheless, the limitations encountered with nucleoside analogues as antiviral chemotherapeutic agents emphasize the need for antiviral drugs which exert their inhibitory activity by different mechanisms of action.

One alternative antiviral drug target is the HSV-encoded ribonucleotide reductase. This enzyme catalyzes the reduction of ribonucleoside diphosphates to the corresponding deoxyribose counterparts[10]. The ribonucleotide reductase of HSV-1 is comprised of two dissimilar subunits. The larger subunit is a homodimer of a 140 kDa polypeptide whereas the smaller subunit is a homodimer of a 38 kDa polypeptide. These subunits associate to form a tight complex of the $\alpha_2\beta_2$ type[11]. The large subunit contains redox-active thiol groups and the small subunit contains a binuclear iron III center and a stable tyrosyl-free radical which are essential for enzymatic activity[12]. These general features of the viral enzyme are similar to those of mammalian ribonucleotide reductases. The HSV enzyme differs, however, biochemically from its mammalian counterpart in that it lacks an absolute requirement for Mg^{2+} and ATP[13,14]. In addition, and most importantly, the viral ribonucleotide reductase is insensitive to feedback inhibition by dATP and dTTP[12] which results in unrestricted synthesis of deoxyribonucleoside triphosphates in a virus infected cell. This apparent requirement for large amounts of deoxyribonucleotides strongly suggests that HSV ribonucleotide reductase may be essential for virus reproduction. Indeed, several reports have suggested that viral ribonucleotide reductase may be required for virus growth in tissue culture and recent studies have clearly demonstrated that selective inhibition of HSV ribonucleotide reductase did correlate with some inhibition of HSV replication[15,16,17]. Furthermore, HSV insertion mutants lacking most of the large subunit of ribonucleotide reductase or a temperature sensitive mutant with a lesion in the gene encoding this subunit produced lower yields of virus progeny in exponentially growing tissue culture cells as compared to wild type virus[18,19]. In resting cells growth of mutant virus was more significantly impaired. Other studies using mutant viruses with alterations in either the large or small subunit of ribonucleotide reductase have shown that HSV ribonucleotide reductase appears to be required for reactivation from latency and essential for pathogenicity in mice[20,21,22]. The only reported exception is a lack of essentiality for pathogenicity in the guinea pig skin model[23]. Taken together, these findings suggest that herpesvirus-specified ribonucleotide reductase may constitute a valid antiviral target.

HSV ribonucleotide reductases are subject to inhibition by a variety of compounds such as nucleotide analogues, metal chelators and radical scavengers[24]. These compounds, however, may lack the required selectivity to inhibit specifically the viral ribonucleotide reductase and may, therefore, turn out not to be very useful in the development of antiherpetic agents. Recently, however, investigators from two laboratories, independently reported that the synthetic nonapeptide H-Tyr-Ala-Gly-Ala-Val-Val-Asn-Asp-Leu-OH, corresponding to the carboxy-terminal nine amino acids of the HSV-1 ribonucleotide reductase small subunit, inhibited HSV-1 ribonucleotide reductase[25,26]. Interestingly, the same peptide did not inhibit mammalian ribonucleotide reductase. It was proposed that the carboxy-terminal nine amino acids of the small subunit provided a binding domain for its association with the large subunit to yield active ribonucleotide reductase holoenzyme. The synthetic nonapeptide inhibited enzymatic activity by competing with the small subunit for binding to the large subunit, thereby preventing the formation of active complexe. These findings suggested that inhibitors whose mechanisms of action would involve disruption of the interaction between the two ribonucleotide reductase subunits may prove useful as selective agents in the treatment of infections caused by herpes simplex viruses. In addition, recent comparisons of the carboxy-terminal amino acid sequences of different ribonucleotide reductase small subunits, and experimental evidence from studies using anti-nonapeptide antibodies have revealed extensive conservation of the carboxyl termini of small subunits from various herpesviruses including HSV-1, HSV-2, VZV, EBV, pseudorabies virus and equine herpesvirus[25,27,28,29,30]. Thus, it could be proposed that the HSV carboxy-terminal nonapeptide and related compounds may inhibit a wide spectrum of herpesviruses. In addition, the pronounced species variation of this same carboxy-terminal region of the small subunit indicates that selectivity may be achieved and that it may be possible to use a similar strategy to design selective inhibitors of not only herpesvirus ribonucleotide reductase, but also bacterial and mammalian ribonucleotide reductases[31,32] for the development of antibacterial and antineoplastic agents.

RESULTS AND DISCUSSION

The purpose of this study was to further investigate the molecular mechanism of inhibition of HSV-1 ribonucleotide reductase by synthetic peptides derived from the carboxy-terminal region of its small subunit in order to develop more potent analogues that could serve as potential antiviral agents. HSV ribonucleotide reductase holoenzyme was

Table I **Structure and IC_{50} values of the nonapeptides tested against HSV-1 ribonucleotide reductase**

Designation	Sequence	IC_{50} (μM)[a]
BILD0006	H-Tyr-Ala-Gly-Ala-Val-Val-Asn-Asp-Leu-OH	33
BILD0013	H-Tyr-Ala-Gly-Thr-Val-Ile-Asn-Asp-Leu-OH	10
BILD0014	Ac-Tyr-Ala-Gly-Thr-Val-Ile-Asn-Asp-Leu-OH	4

[a] IC_{50} values were estimated from the dose-response curves shown in Figure 1 using SAS software (SAS Institute, Cary, North Carolina) in which non-linear regression analysis using the Hill equation[34] was applied to the percent inhibition-concentration data.

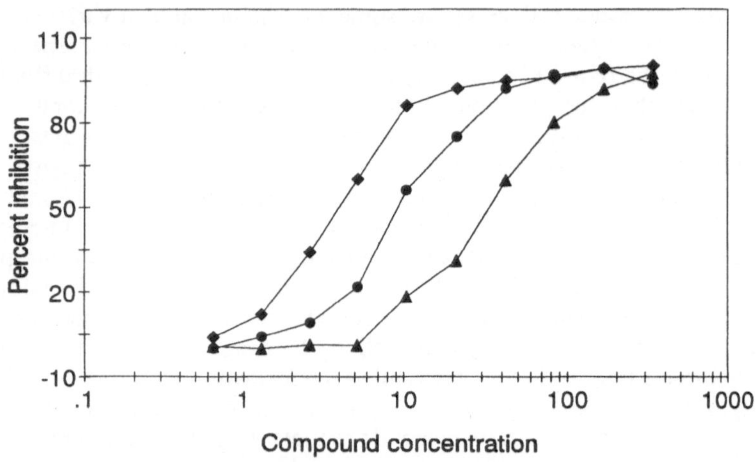

Figure 1 **Dose-dependent inhibition of ribonucleotide reductase isolated from HSV-1-infected BHK-21 cells.** Aliquots of the viral enzyme preparation (0.03 units) were incubated with increasing concentrations of BILD0006 (▲), BILD0013 (●) and BILD0014 (◆), and residual ribonucleotide reductase activity was measured. In these experiments the CDP concentration in the assay mixture was 54 µM. After 30 minutes incubation at 37°C, the reactions were stopped by adding 2.5 units of calf alkaline phosphatase (Boehringer Mannheim) which converted the nucleotides to the corresponding nucleosides. The resulting cytidine and deoxycytidine were separated by thin layer chromatography on PEI-cellulose and quantitated using a radioanalytical imaging system (Ambis Inc., San Diego, CA). The results, presented as the percent inhibition, were calculated from residual enzymatic activity which was expressed as a percentage of control activity determined in the absence of inhibitor.

partially purified from HSV-1 strain F-infected baby hamster kidney (BHK)-21 cells as described elsewhere[33]. The holoenzyme preparation used in this study had a specific activity of 0.44 units/mg protein, as determined by monitoring the conversion of [^{14}C] CDP to [^{14}C] dCDP according to a published procedure[31], except that $MgCl_2$ and ATP were omitted from the assay mixture. One unit of reductase was defined as that amount of enzyme which reduced 1 nanomole CDP per minute. The nonapeptides BILD0006 and BILD0013, corresponding respectively to the last nine amino acids of the ribonucleotide reductase small subunit of HSV-1 and VZV, and the nonapeptide BILD0014, a N-terminal acetylated form of BILD0013, were all prepared by standard solid-phase peptide synthesis (see Table I for structures). The effect of these three nonapeptides on the activity of ribonucleotide reductase isolated from HSV-1-infected BHK-21 cells is shown in Figure 1. Each peptide was tested at two-fold serial dilutions and the percent inhibition was calculated from residual enzymatic activity. Non-linear regression analysis based on the Hill equation[34], was employed to determine the inhibitor concentration that yielded 50% of the maximal inhibition (i.e., IC_{50}). Results presented in Figure 1 and Table I clearly indicated that all three peptides were relatively potent inhibitors of the viral reductase. Interestingly, the nonapeptide BILD0013 derived from the VZV ribonucleotide reductase small subunit sequence was a better inhibitor of the HSV-1 enzyme than the nonapeptide BILD0006 corresponding to the carboxy-terminus of the HSV-1 ribonucleotide reductase small subunit. In addition, N-terminal acetylation of BILD0013 yielded a better inhibitor relative to its unacetylated analogue. This result suggested that acetylation at the N-terminus produced a peptide analogue with increased affinity for the ribonucleotide reductase large subunit.

Several investigators have derived simple mathematical equations which utilize the determination of IC_{50} values at several levels of substrate concentrations with which to distinguish between different mechanisms of inhibition[35,36]. Thus, to determine the type of ribonucleotide reductase inhibition by the nonapeptides shown in Table I, the IC_{50} of BILD0014 was measured at various CDP concentrations. In order to validate a classical substrate competitive inhibition profile, experiments were performed in parallel with ADP, a known competitive inhibitor of ribonucleotide reductase with respect to CDP reduction[24]. For classical substrate competitive inhibition,

$$IC_{50} = Ki + Ki/Km\ S,$$

where Ki, Km and S represent the inhibition constant, the Michaelis-Menten constant and the substrate concentration, respectively. A plot of IC_{50} versus S should give a straight line with a IC_{50}-intercept equal to Ki and a slope equal to Ki/Km. For classical non-competitive inhibition,

$$IC_{50} = Ki$$

and therefore, IC_{50} should be independent of the substrate concentration. As is clearly demonstrated in Table II, the IC_{50} of BILD0014 was the same within experimental error, regardless of the substrate concentration used in the assay, suggesting that BILD0014, like the nonapeptide BILD0006[26] acted as a non-competitive inhibitor with respect to CDP. In

Table II **Effect of CDP concentration on the inhibition of HSV-1 ribonucleotide reductase by BILD0014 and adenosine-5'-diphosphate (ADP)**

	IC_{50} (μM)[a]	
CDP (μM)	**BILD0014**	**ADP**
13.5	3.8	209
27.0	3.9	480
54.0	3.9	623
81.0	4.0	1028

[a] IC_{50} values for BILD0014 and for ADP were determined as outlined in the legends to Figure 1 and Table I in the presence of the indicated CDP concentrations. Each IC_{50} value represents the mean of two determinations. Variation between individual measurements was less than 20%.

Figure 2 **Immunoprecipitation of HSV-1 ribonucleotide reductase with monoclonal antibody 535 and dissociation of the subunits by the nonapeptide BILD0006**. Immunoprecipitations were performed with 535 antibody after incubation of ribonucleotide reductase holoenzyme in the presence (lane 1) or absence (lane 2) of BILD0006 as outlined in the text. An aliquot of enzyme (50 µg) was analyzed under the same conditions without immunoprecipitation (lane 3). The Western blot was performed using a mixture of two monoclonal antibodies, 535 and 932, that recognize the small and the large subunits, respectively. The molecular weight corresponding to the large (140 kDa) and small (38 kDa) ribonucleotide reductase subunits are indicated by arrows.

marked contrast, and as expected, ADP clearly behaved as a competitive inhibitor of the substrate for the viral enzyme.

To investigate whether the nonapeptide was capable of dissociating ribonucleotide reductase large and small subunits, immunoprecipitation experiments were performed using the monoclonal antibody 535 which was raised against the small subunit of HSV-1 ribonucleotide reductase[11]. This monoclonal antibody was chosen since it was demonstrated to be non neutralizing and to bind the intact holoenzyme comprising both the large and the small subunits. Partially purified ribonucleotide reductase from HSV-1-infected cells (300 µg) in 50 mM HEPES, pH 7.5, 2 mM DTT was incubated on ice in the presence or absence of 1 mM BILD0006 for 30 minutes in a total volume of 500 µL. Approximately 25 µL of sedimented antibody 535-Sepharose 4B, prepared as described previously[11], were then added and the reaction mixtures (total volume, 600 µL) were incubated for 3 additional hours at 4°C. The Sepharose was then pelleted by centrifugation at 10,000 rpm for 5 minutes and washed with 1 mL 50 mM HEPES, pH 7.5, 0.5 M KCl to remove non-specifically bound proteins. The pelleted proteins were resuspended in Laemmli buffer, separated on a 5-20% linear gradient SDS polyacrylamide gel and transferred to a nitrocellulose membrane as described previously[11]. The membranes were washed with 50 mM Tris-HCl, pH 7.6, 0.5 M NaCl and 0.5% Tween 20 and then incubated for 3 hours at room temperature with a mixture of antibodies containing 10 μg/mL each of 535 and 932 antibody. The 932 antibody, a non neutralizing monoclonal antibody, was directed against the large subunit of HSV ribonucleotide reductase. After washing, bound antibodies were detected with rabbit anti-mouse antibodies conjugated to alkaline phosphatase. Figure 2 shows that in the absence of BILD0006, the 535 antibody was capable of co-immunoprecipitating both the small and the large subunits. However, after preincubation of the holoenzyme with the nonapeptide BILD0006, only the small subunit was detected on the Western blot, implying that the nonapeptide disrupted the holoenzyme complex and prevented the reassociation of the two ribonucleotide reductase subunits. These results support the model proposed by others[25,26,37] that the inhibitory nonapeptide competed with the small subunit for binding to the large subunit, and that loss of enzymatic activity is directly related to separation of ribonucleotide reductase subunits.

In conclusion, our results confirm and extend previous findings by others[25,26] on the mechanism of inhibition by the nonapeptide and related compounds. The results presented here support the concept for the development of antiherpetic agents based on dissociation of ribonucleotide reductase subunits. Moreover, these studies provide the basis from which synthetic peptides derived from amino acid sequences involved in protein-protein interaction may be designed to interfere specifically with essential enzymatic activities. Efforts in our laboratories are presently directed towards assessing the potential antiviral activity of such peptidic inhibitors. Since it has already been observed that incubation of HSV infected cells with increasing concentrations of the authentic nonapeptide had no effect on the yield of infectious virus[25,38,39,40], we are now attempting to design peptides and peptidomimetics with greater potency, stability and cell penetration capability. In this regard, experiments are currently in progress to determine the molecular configuration of the inhibitory peptides and to elucidate the primary amino acid sequence of the large subunit that interacts specifically with the carboxyl terminus of the small subunit of HSV-1 ribonucleotide reductase. This information will be most valuable in the future design of more efficacious inhibitors.

ACKNOWLEDGEMENTS

We wish to thank R. Plante for synthesizing the inhibitory nonapeptides, L. Thelander for procuring 535 and 932 monoclonal antibodies and R. Krogsrud, R. Déziel, A.-M. Bonneau, G. Cosentino, N. Moss and J.G. Chafouleas for helpful comments on the manuscript. The excellent secretarial support of C. Damato is also acknowledged.

* To whom correspondence should be addressed.

REFERENCES

1. B. N. Fields, "Virology", Raven Press, New York (1990).
2. J.G. Stevens, Human herpesviruses: a consideration of the latent state, Microbiol. Rev. 53: 18 (1989).
3. G.J. Galasso, R.J. Whitley, and T.C. Merigan, "Antiviral Agents and Viral Diseases of Man", Raven Press, New York (1990).
4. D.J. McGeoch, M.A. Dalrymple, A.J. Davison, A. Dolan, M.C. Frame, D. McNab, L.J. Perry, J.E. Scott, and P. Taylor, The complete DNA sequence of the long unique region in the genome of herpes simplex virus type 1, J. Gen. Virol. 69:1531 (1988).

5. R.W. Honess, and B. Roizman, Regulation of herpesviruses macromolecular synthesis: I. Cascade regulation of the synthesis of three groups of viral proteins, J. Virol. 14:8 (1974).
6. W.M. Shannon, Mechanisms of action and pharmacology: chemical agents, in: "Antiviral Agents: Viral Diseases of Man", G.J. Galasso, R.J. Whitley, and T.C. Merigan, eds., Raven Press, New York (1984).
7. S.A. Plotkin, S.E. Starr, and C.K. Bryan, In vivo and in vitro responses of cytomegalovirus to acyclovir, Am. J. Med. 73:257 (1982).
8. E. DeClercq, J. Descamps, P. DeSomer, P.J. Barr, A.S. Jones, and R.T. Walker, (E)-5-(2-Bromovinyl)-2'-deoxyuridine: a potent and selective antiherpes agent, Proc. Natl. Acad. Sci. U.S.A. 76:2947 (1979).
9. H.J. Field, The development of antiviral drug resistance, in: "Antiviral Agents: The Development and Assessment of Antiviral Chemotherapy", H.J. Field, ed., CRC Press, Boca Raton, Florida (1988).
10. P. Reichard, Interactions between deoxyribonucleotide and DNA synthesis, Annu. Rev. Biochem. 57:349 (1988).
11. R. Ingemarson, and H. Lankinen, The herpes simplex virus type 1 ribonucleotide reductase is a tight complex of the type composed of 40 K and 140 K proteins, of which the latter shows multiple forms due to proteolysis, J. Virol. 156:417 (1987).
12. M. Lammers, and H. Follman, The ribonucleotide reductases: a unique group of metalloenzymes essential for cell proliferation, Struct. Bonding 54: 27(1983).
13. D. Huszar, and S. Bachetti, Partial purification and characterization of the ribonucleotide reductase induced by herpes simplex virus infection of mammalian cells, J. Virol. 37:580 (1981).
14. Y. Langelier, M. Deschamps, and G. Buttier, Analysis of dCMP deaminase and CDP reductase levels in hamster cells infected with herpes simplex virus, J. Virol. 26:547 (1978).
15. L.M. Nutter, S.P. Grill, and Y.-C. Cheng, Can ribonucleotide reductase be considered as an effective target for developing antiherpes simplex virus type II (HSV-2) compounds? Biochem. Pharmac. 34:777 (1985).
16. T. Spector, D.R. Averett, D.J. Nelson, C.U. Lambe, R.W. Morrison, M.H. StClair, and P.A. Furman, Potentiation of antiherpetic activity of acyclovir by ribonucleotide reductase inhibition, Proc. Natl. Acad. Sci. USA 82:4254 (1985).
17. S.R. Turk, C. Shipman, and J.C. Drach, Selective inhibition of herpes virus ribonucleotide diphosphate reductase by derivatives of 2-acetylpyridine thiosemicarbazone, Biochem. Pharmac. 35:1539 (1986).
18. V.G. Preston, J.W. Palfreyman, and B.M. Dutia, Identification of a herpes simplex virus type 1 polypeptide which is a component of the virus-induced ribonucleotide reductase, J. Gen. Virol. 65:1457 (1984).
19. D.J. Goldstein, and S.K. Weller, Herpes simplex virus type 1-induced ribonucleotide reductase activity is dispensable for virus growth and DNA synthesis: isolation and characterization of an ICP6 *lac* Z insertion mutant, J. Virol. 62:196 (1988).
20. J.M. Cameron, I. McDougall, H.S. Marsden, V.G. Preston, M.D. Ryan, and J.H. Subak-Sharpe, Ribonucleotide reductase encoded by herpes simplex virus is a determinant of the pathogenicity of the virus in mice and a valid antiviral target, J. Gen. Virol. 69:2607 (1988).
21. J.G. Jacobson, D.A. Leib, D.J. Goldstein, C.L. Bogard, P.A. Schaffer, S.K. Weller, and D.M. Coen, A herpes simplex virus ribonucleotide reductase deletion mutant is defective for productive acute and reactivatable latent infections of mice and for replication in mouse cells, Virology 173:276 (1989).

22. R.T. Kintner, C.R. Brandt, R.J. Visally, A.M. Pumfery, and D.R. Grau, The herpes simplex virus ribonucleotide reductase is required for ocular virulence, Invest. Ophtalmol. Visual Sci. 32:852 (1991).
23. S.R. Turk, N.A. Kik, G.M. Birch, D.. Chiego, Jr., and C. Shipman, Jr., Herpes simplex virus type 1 ribonucleotide reductase null mutants induce lesions in guinea pigs, Virology 173:733 (1989).
24. T. Spector, Ribonucleotide reductase encoded by herpesviruses: inhibitors and chemotherapeutic considerations, in: "International Encyclopedia of Pharmacology and Therapeutics", J.G. Cory, and A.H. Cory, eds., Pergamon Press, Elmsford, New York (1989).
25. B.M. Dutia, M.C. Frame, J.H. Subak-Sharpe, W.N. Clark, and H.S. Marsden, Specific inhibition of herpes ribonucleotide reductase by synthetic peptides, Nature (London) 321:439 (1986).
26. E.A. Cohen, P. Gaudreau, P. Brazeau, and Y. Langelier, Specific inhibition of herpesvirus ribonucleotide reductase by a nonapeptide derived from the carboxy terminus of subunit 2, Nature (London) 321:441 (1986).
27. A.J. Davison and J.E. Scott, The complete DNA sequence of varicella-zoster virus, J. Gen. Virol. 67:1759 (1986).
28. R. Baer, A.T. Bankier, M.D. Biggin, P.L. Deininger, P.J. Farewell, T.J. Gibson, G. Hatfull, G.S. Hudson, S.C. Satchwell, C. Séguin, P.S. Tuffnell, and B.G. Barrell, DNA sequence and expression of the B95-8 Epstein-Barr virus genome, Nature 310:207 (1984).
29. E. Telford, H. Lankinen, and H.S. Marsden, Inhibition of equine herpesvirus type 1 subtype 1-induced ribonucleotide reductase by the nonapeptide YAGAVVNDL, J. Gen. Virol. 71:1373 (1990).
30. E.A. Cohen, H. Paradis, P. Gaudreau, P. Brazeau, and Y. Langelier, Identification of viral polypeptides involved in pseudorabies virus ribonucleotide reductase activity, J. Virol. 61:2046 (1987).
31. G. Cosentino, P. Lavallée, S. Rakhit, R. Plante, Y. Gaudette, C. Lawetz, P. Whitehead, J.-S. Duceppe, C. Lépine-Frenette, N. Dansereau, C. Guilbeault, Y. Langelier, P. Gaudreau, L. Thelander, and Y. Guindon, Specific inhibition of ribonucleotide reductases by peptides corresponding to the C-terminal of their second subunit, Biochem. Cell Biol. 67:79 (1991).
32. F.-D. Yang, R.A. Spanevello, I. Celiker, R. Hirschmann, H. Rubin, and B.S. Cooperman, The carboxyl terminus heptapeptide of the R2 subunit of mammalian ribonucleotide reductase inhibits enzyme activity and can be used to purify the R1 subunit, FEBS Lett. 272:61 (1990).
33. E.A. Cohen, J. Charron, J. Perret and Y. Langelier, Herpes simplex virus ribonucleotide reductase induced in infected BHK-21/C13 cells: biochemical evidence for the existence of two non-identical subunits, H1 and H2, J. Gen. Virol. 66:733 (1985).
34. M. Dixon and E.C. Webb, "Enzymes", Academic Press, New York (1979).
35. Y.-C. Cheng, and W.H. Prusoff, Relationship between inhibition constant (Ki) and the concentration of inhibitor which causes 50 percent inhibition (I50) of an enzymatic reaction, Biochem. Parmac. 22:3099 (1973).
36. P.J.F. Henderson, A linear equation that describes the steady-state kinetics of enzymes and subcellular particles interacting with tightly bound inhibitors, Biochem. J. 127:321 (1972).
37. H. Paradis, P. Gaudreau, P. Brazeau, and Y. Langelier, Mechanism of inhibition of herpes simplex virus (HSV) ribonucleotide reductase by a nonapeptide corresponding to the carboxyl terminus of its subunit 2. Specific binding of a photoaffinity analog, [4'-azido-Phe^6] HSV2-(6-15), to subunit 1, J. Biol. Chem. 263:16045 (1988).

38. H. Paradis, Y. Langelier, J. Michaud, P. Brazeau, and P. Gaudreau, Studies on *in vitro* proteolytic sensitivity of peptides inhibiting herpes simplex virus ribonucleotide reductases lead to discovery of a stable and potent inhibitor, Int. J. Peptide Protein Res. 37:72 (1991).
39. E. Telford, A. Owsianka, and H.S. Marsden, Stability of the herpesvirus ribonucleotide reductase-inhibiting nonapeptide YAGAVVNDL in extracts of HSV1-infected cells, Antiviral Chem. Chemother. 1:223 (1990).
40. Our unpublished results.

ANTIVIRAL EFFECTS OF HERPES SIMPLEX VIRUS SPECIFIC ANTI-SENSE NUCLEIC ACIDS

Edouard M. Cantin, Gregory Podsakoff,
Dru E. Willey, and Harry Openshaw

City of Hope National Medical Center
Department of Neurology
Duarte, CA 91010

SUMMARY

We have targeted mRNA sequences encompassing the translation initiation codon of the essential herpes simplex virus type 1 (HSV-1) IE3 gene with three kinds of anti-sense molecule. Addition of a 15mer oligodeoxyribonucleoside methylphosphonate to tissue culture cells resulted in suppression of viral replication. HSV-1 replication was also inhibited in cultured cells containing anti-sense vectors expressing transcripts complementary to the IE3 mRNA. We have also constructed a ribozyme which upon base pairing with the target IE3 mRNA induces cleavage at the predicted GUC site. A major obstacle to anti-sense studies in animals is drug delivery of preformed anti-sense molecules to ganglionic neurons, the site of HSV latency and reactivation. We speculate as to how this may be accomplished through carrier compounds which are taken up by nerve terminals and transported by retrograde axoplasmic flow. By the same route, HSV itself may be used as an anti-sense vector.

INTRODUCTION

The human herpesviruses share a common ultrastructural morphology, certain aspects of replication and the ability to establish a lifelong latent infection, as best exemplified by herpes simplex virus type 1 and 2 (HSV) (Roizman, 1983; Wildy et al., 1982). HSV causes a wide spectrum of disease, ranging from relatively mild primary lesions involving the skin, mucous membranes and corneal epithelium, to severe and often fatal episodes of encephalitis. Periodic reactivation of latent virus over the lifetime of the host results in recurrent skin lesions, often associated with significant morbidity (Corey and Spear, 1986).

At the time of initial contact, HSV replicates in epithelial cells at the primary site of infection, eventually spreading to invade sensory nerve terminals. The virus, in the form of a sub-viral particle or perhaps as a nucleoprotein complex, spreads in the intra-axonal space to the nerve cell bodies in sensory ganglia where a latent infection is established (Stevens, 1989; Roizman and Sears, 1987). During latency, the viral genome is essentially quiescent (Cantin, et al., 1987), except for the production of two or three latency associated transcripts (LATs) which accumulate to high levels in the neuronal nucleus (Stevens et al., 1987; Puga and Notkins, 1987). Studies of LAT deletion mutants suggest that these transcripts, which are complementary or anti-sense to the ICPO mRNA, are dispensable for establishment or maintenance of latency, but appear important for efficient reactivation (Leib et al., 1989a; Steiner et al, 1989). There is however, no evidence to support an anti-sense mechanism of action for LAT despite data showing a requirement for ICPO during reactivation (Leib et al, 1989b).

In this report, we review briefly the limitations of current anti-HSV therapy, vaccines and anti-viral drugs, and discuss the general concept of anti-sense DNA and RNA as potential anti-viral agents. We present results of anti-sense inhibition of HSV replication in tissue culture cells and speculate as to how anti-sense may be delivered to ganglionic neurons and applied as anti-HSV therapy.

CURRENT ANTI-HSV TREATMENT STRATEGIES

HSV vaccines for human use are still in the developmental stage. Vaccination is intended mainly for seronegative individuals at risk for exposure, but it is not enough for a vaccine to prevent clinical disease. To be fully effective, viral colonization of the ganglia must be blocked, thereby eliminating both recurrences and asymptomatic shedding in the vaccinee with transmission to contacts. This has been accomplished in experimental animals using subunit preparations and recombinant viral vaccines (Meignier et al., 1987; Stanberry et al., 1987; Cremer et al., 1985; Willey et al, 1988). Since HSV DNA can produce neoplastic transformation (Galloway and McDougall, 1983), it is unlikely that a live HSV vaccine will be approved for human use in the near future. Whether subunit preparations shown to work in animals will be similarly effective in humans is uncertain. Latency can be established in the absence of viral replication in nervous system cells (Coen et al., 1989; Dobson et al, 1990); therefore, to prevent latency the immune system must neutralize input virus at the skin before invasion of sensory nerve terminals. Once in the intra-axonal space, virions are sheltered from immunity.

The effectiveness of anti-viral drugs is well established, especially for acyclovir (ACV) and particularly in the treatment of primary HSV infections (Crumpacker, 1989). In immunocompromised patients, recurrent herpetic infections run a protracted course, and ACV is useful in this context; whereas, only a modest benefit has been achieved for recurrences suffered by immunocompetent individuals (Spruance et al., 1990). Recurrence can be prevented, however by chronic ACV therapy (Straus et al., 1984); although such prophylaxis is generally reserved for individuals with very frequent eruptions and chronic therapy does not eliminate latent virus. Only occasionally have ACV-resistant mutants been documented during prophylaxis of immunocompetent patients (Nusinoff et al., 1986), but ACV-resistant mutants are a problem in AIDS and other immunosuppressive disorders (Erlich et al., 1989; Hirsch and Schooley, 1989). Viral thymidine kinase (tk) is required for phosphorylation of ACV to the active form and most of the ACV-resistant mutants are tk-deficient, although rare mutants with altered tk substrate-specificity or altered DNA polymerase have been isolated (Ellis et al., 1987; Parker et al, 1987). The tk-deficient mutants isolated from clinical material replicate poorly or not at all in the nervous system; and although defined tk deletion mutants can establish latency, they are reactivation-incompetent in explant culture (Coen et al, 1989). Foscarnate, a pyrophosphate analogue that inhibits the viral DNA polymerase, has been used to treat ACV-resistant mucocutaneous lesions in AIDS (Chatis et al., 1989).

Unlike vaccines or nucleoside analogue drugs, anti-sense molecules can selectively affect viral gene expression, and therefore anti-sense strategies have the potential of molecularly protecting cells from latency or from reactivation.

ANTI-SENSE INHIBITION OF GENE EXPRESSION

In retrospect, the first application of anti-sense was the technique of hybrid arrest translation used to map viral genomic coding sequences (Paterson et al., 1977). In this technique, complementary DNA sequences inhibit *in vitro* translation by forming an RNA-DNA duplex through Watson-Crick base pairing. Later, Zamencnik and Stephenson (1978) in a pioneering study, demonstrated the inhibition of Rous sarcoma virus replication by adding to the tissue culture cell media an oligonucleotide 30 bases long, complementary to the 5' and 3' reiterated terminal sequence of the virus.

Rapid advances in synthetic nucleic acid chemistry have subsequently resulted in the development of modified oligonucleotides with improved cell penetration and enhanced nuclease resistance. The commonest modifications have involved substitution of the oxygen atom of the internucleotide phosphodiester linkage with some substituent such as a methyl group, giving an oligonucleotide methylphosphonate. Modified oligonucleotides, including those with reactive groups such as psoralen or acridine orange at their ends are being increasingly used to modulate gene expression in diverse systems (Uhlmann and Peyman, 1990). With the recognition that anti-sense RNA functions naturally to regulate gene expression in prokaryotes (Coleman et al., 1984), came the development of RNA based anti-sense expression vector systems which have been used to express various prokaryotic genes and also to inhibit bacteriophage replication (Coleman et al., 1985). Izant and Weintraub (1984), were first to use an anti-sense gene in eukaryotic cells, to block expression of the HSV thymidine kinase gene, a process they termed "anti-sense inhibition." Since then, anti-sense has been used to inhibit gene expression in diverse systems ranging from frog oocytes to transgenic mice. Compared to oligonucleotides, anti-sense RNA offers the significant advantage of permanently altering the phenotype of a cell, when the anti-sense gene is integrated into the genome and constitutively expressed. This has been clearly demonstrated in plant systems in which the anti-sense phenotype has been shown to be heritable (Krol et al., 1988a, 1988b).

An exciting and important recent discovery is that RNA, besides being an informational macromolecule, can participate in autocatalytic cleavage reactions and therefore can be regarded as possessing enzymatic activity. RNA enzymes or ribozymes are involved in diverse biochemical processes ranging from the self-splicing reaction of the large, ribosomal RNA precursor of Tetrahymena to resolution of RNA replication intermediates of plant viroid and virusoid RNAs (Cech and Bass, 1986; Hutchins et al., 1986). Studies, particularly of the viroid-virusoid ribozymes have identified 13 conserved nucleotides involved in extensive secondary structures, essential to formation of the active, catalytic "hammerhead" ribozyme configuration (Buzayan et al., 1986; Forster and Symons, 1987). Strong impetus for the development of transacting ribozymes as potential therapeutic agents came from the demonstration that a catalytically active, hammerhead ribozyme could be formed from two RNAs acting in trans, one possessing the thirteen conserved catalytic domain nucleotides and the other the cleavage sequence NUX (where N is any base and X is A, U, or C) (Ulenbeck, 1987; Haseloff and Gerlach, 1988). Anti-sense DNA and RNA are now being increasingly used to inhibit gene expression in diverse systems ranging from cell free *in vitro* protein synthesis systems to transgenic animals (Krol, 1988b), particular emphasis is on exploiting anti-sense to block expression of deleterious genes, especially viral genes and those of other disease causing pathogens. Anti-sense inhibition is especially attractive as a potential antiviral therapy because of the exquisite specificity conferred by the requirement for specific base pairing interactions between target mRNA and anti-sense DNA (oligonucleotide) or RNA molecule. Statistically, the sequence of a 17-mer oligonucleotide occurs only once in the human genome. Additionally, with anti-sense there is the flexibility of choice of the viral gene to target and the stage of replication at which to intervene; this is not possible when more traditional antiviral drugs are used. The ultimate challenge is to exploit the potential of anti-sense to render specific cell populations in an organism permanently resistant to viral infection, the concept of molecular immunity.

ANTI-SENSE INHIBITION OF HSV-1 REPLICATION

We are exploring the use of anti-sense to inhibit expression of essential HSV regulatory genes, thereby inhibiting viral replication. Though conceptually simple, the successful application of anti-sense depends on several factors, including accessibility of the target mRNA, stability of anti-sense molecule, and delivery not only to the appropriate cell in an animal, but also to the appropriate intracellular compartment. We chose as an anti-sense target the HSV-1 IE3 gene encoding the ICP4 (or Vmw 175) polypeptide, a multifunctional, regulatory protein, the synthesis of which is required continuously for virus replication (Watson and Clements, 1980; DeLuca et al., 1985). In the absence of ICP4, HSV immediate early (IE) polypeptides are produced but the subsequent cascade of viral protein synthesis is abrogated and there is no virion assembly. Moreover, because ICP4 very likely interacts stoichiometrically with other viral IE proteins and with cellular proteins (O'Hare and Hayward, 1985; Knipe, 1989), we reasoned that anti-sense mediated reduction of ICP4 expression may suffice to inhibit HSV-replication, and that a complete block of ICP4 expression not be necessary.

A synthetic oligodeoxyribonucleoside methylphosphonate (OMP) designed to block translation of IE3 mRNA was synthesized and assayed for HSV specific antiviral activity *in vitro*. The 15-mer IE3 OMP is complementary to sequences encompassing the translation initiation codon (ATG), a region of mRNA that is likely to be relatively free of secondary structure and therefore accessible to anti-sense molecules. The same region of IE3 mRNA was also targeted with anti-sense RNA, including a ribozyme designed to cleave at the GUC site two bases downstream from the ATG. This fortuitous sequence arrangement should allow for evaluation of relative efficacy of standard anti-sense compared to catalytic anti-sense or ribozymes. Ts'o and collaborators (Smith et al., 1986; Kulka et al., 1989) have previously reported inhibitory activity for an OMP designed to block splicing of the HSV IE4/5 pre-mRNA. An analogous 15-mer OMP, complementary to sequences around IE4/5 splice acceptor junction, was used as a positive control for our experiments. The genomic location of the HSV-1 IE genes and the IE3 and IE4/5 mRNA sequences targeted with various types of anti-sense molecules is shown in Figure 1.

In Figure 2, the effect of OMP IE3 on HSV-1 replication is compared to OMP IE4/5. Vero cells were incubated with the OMP's for one hour, the OMP's removed and the cells then inoculated with HSV-1 or, as a control for specificity, with vaccina virus at a multiplicity of infection of 0.1 plaque forming unit per cell. At 24 hours after virus inoculation, the cells were harvested and viral titers determined. The results in Figure 2 show that both anti-sense OMP's inhibited HSV-1 replication. OMP IE4/5 was more active than OMP IE3; approximately 4-fold more OMP IE3 being required to achieve the same degree of inhibition. OMP IE4/5 significantly inhibited vaccina virus replication at a concentration of 50 μM; where as, only a slight inhibitory effect of OMP IE3 was noted at a concentration of 100 μM.

A

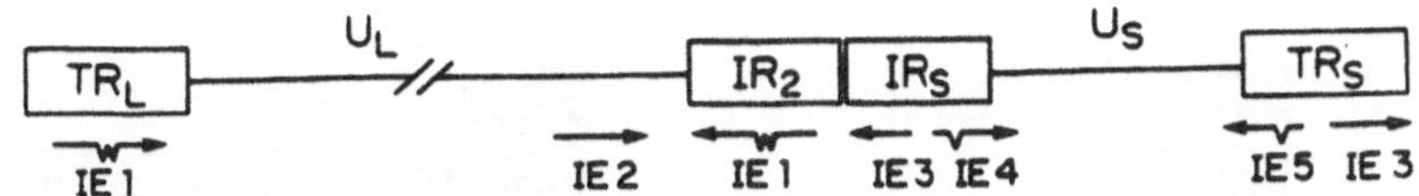

B

INTRON EXON

HSV IE4/5 mRNA: 5'-UUCCCGCAG]GAGGAACGUCCU-3'
3'-GGGCGTC]CTCCTTGC-5' Anti-sense Oligo

HSV IE3 mRNA: 5'-CCGCAUCGGCGAUGGCGUCGG-3'
3'-CGTAGCCGCTACCGC-5' Anti-sense Oligo

HSV IE3 mRNA: 5'-GCG**AUG**GCGUCGGAGAACAAG-3'
3'-CUACCGCA CCUCUUGU-5
A C
A U
G A GGAC G G
U G CCUG A G U A
Ribozyme

Figure 1. A: HSV-1 genome, B: anti-sense OMP's binding to HSV-1 IE4/5 mRNA and HSV-1 IE3 mRNA and ribozyme binding to HSV-1 IE3 mRNA.

The effect of OMP IE4/5 on vaccina virus is probably indirect; vaccina virus replicates in the cytoplasm and does not undergo splicing. Moreover, the inhibitory effect of OMP IE4/5 on HSV replication is difficult to explain since, unlike IE3, IE4/5 are dispensable genes for lytic HSV replication in cultured cells (Knipe, 1989). These results emphasize the need for carefully chosen controls of specificity. Use of the corresponding sense oligonucleotide is not a sufficient control, and attempts to measure non-specific effects of oligonucleotides on cellular physiology suffer from a lack of sensitive assays. The ideal control is a modified virus in which the target sequence has been deliberately mutated to preclude binding of the anti-sense species. A more reasonable control however, is the use of a related virus such as vaccinia virus, as reported here.

We have also constructed IE3 anti-sense expression vectors. In brief, a 1.2 kB BamH1/Nru1 fragment containing the 5' end of the IE3 gene, but lacking the promoter and transcription start site was isolated and inserted in the anti-sense orientation downstream of various promoters. The SV40 early polyadenylation signal was inserted 3' of the IE3 gene fragment completing the anti-sense transcription unit. These anti-sense vectors also contained a dominant selectable marker gene specifying resistance to either G418 (Neo^r gene) or mycophenolic acid and HAT (*E. coli* gpt gene) in eukaryotic cells. Anti-sense transcripts expressed from these vectors are complementary to the IE3 mRNA sequence extending from +31 to +1222, relative to the transcription start site at +1. IE3 anti-sense vectors were introduced into NIH 3T3 cells, using lipofectin reagent (GIBCO-BRL) and approximately 100 clonal anti-sense lines were isolated by growth in selective media. Prior to screening, the cells were passaged 3-4 times in non-selective media. The anti-sense lines were initially screened for resistance to HSV-1 by plaque reduction assay and promising clones (those having smaller plaques or fewer plaques) were assayed for virus growth.

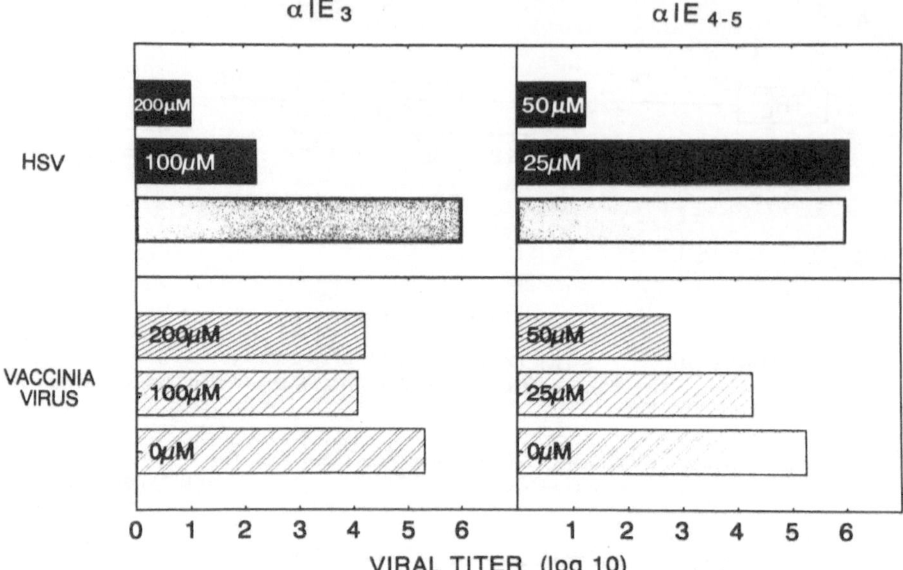

Figure 2. Effect of anti-sense OMP's to IE3 (alpha IE3) and IE4/5 (alpha IE4-5) on HSV-1 and vaccinia virus replication in Vero cells.

Selected cell lines were inoculated with HSV-1 F strain or, as a control for specificity, vaccinia virus strain WR, at a multiplicity of infection of 0.1 plaque forming unit per cell; and 24 hours virus growth yield was determined by plaque assay. As shown in Figure 3, several lines derived from either the RSV-LTR promoter or the HSV IE3 promoter (aS) constructs, reduced HSV replication by 3-4 logs as compared to infection of the same cells with vaccinia virus, or to infection of normal NIH 3T3 cells. Some lines were also slightly inhibitory for vaccinia virus growth and the reason for this is not known. Upon continued passage in non-selective media, these cells rapidly lost their ability to block HSV replication and after about passage 20, no inhibitory effect was evident. Whether this is due to loss of anti-sense sequences or to silencing of expression has not yet been determined.

Finally, we have designed an IE3 specific transacting ribozyme with 8 bases of flanking sequence mediating base pairing to the IE3 RNA. This ribozyme cleaved an IE3 transcript precisely at the predicted GUC site (Figure 1); but it did not turn over multiple substrate molecules, as expected of a catalytic ribozyme. This catalytic failure appears to be due either to feedback inhibition by the cleavage products or to failure of the cleavage products to dissociate for the ribozyme. The first possibility can be evaluated by determining whether the cleavage products specifically associate with the ribozyme or substrate under cleavage conditions. To evaluate the second possibility, ribozymes with shorter base pairing sequences can be tested. Because of the high G+C content of the HSV genome, and the IE3 gene in particular, it may be impossible to design ribozymes which retain both specificity and catalytic capability since these competing attributes are both a function of the length of the flanking, base pairing sequences. Comparative studies of antiviral efficacy of standard anti-sense and ribozyme targeted to the same sequence (Figure 1) will allow rational design of appropriate anti-sense molecules (standard or catalytic) for use in future animal studies.

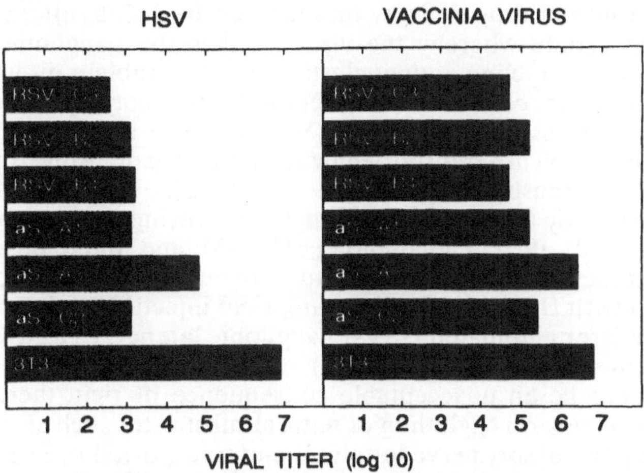

Figure 3. HSV-1 and vaccinia virus replication in normal NIH 3T3 cells (3T3) and 3T3 cells transfected with IE3 anti-sense genes expressed from the Rous sarcoma virus LTR (RSV) or the HSV-1 IE3 (aS) promoter. Cells were inoculated with a multiplicity of infection of 0.1 plaque forming units per cell, and the virus yield was determined in 24 hours. The various anti-sense clonal cell lines tested were designated C3, B2, etc.

In summary, there is now ample evidence that anti-sense can effectively block HSV replication in tissue culture systems. Preliminary studies in the mouse have also shown that "Matagen" (OMP IE4/5) can block HSV skin lesion formation and reduce ganglionic viral titers, when applied topically (Ts'o et al., 1987). The prospect of using anti-sense as a therapeutic agent for combating viral diseases in humans is an exciting challenge being taken up by several laboratories. Currently, the technology is in its infancy; there are a great many problems to overcome before anti-sense can be applied effectively in an animal.

ANIMAL STUDIES WITH ANTI-SENSE

Perhaps the greatest obstacle is the development of a drug delivery system for expression vectors and/or preformed anti-sense molecules (PASMS, i.e. stabilized oligodeoxyribonucleotides or ribozymes). The drug delivery system must provide access to the sites of HSV infection: the skin, eye, and especially sensory ganglia. It should be possible to deliver PASMS to the skin using standard topical methods, and treatment of eye infections may be facilitated with collagen eye shields impregnated with PASMS. We reported that ACV-containing collagen eye shields resulted in reduced HSV titers in the eye and trigeminal ganglion at 4 days after HSV-1 inoculation by the ocular route (Willey et al., 1991). The targeted delivery of PASMS to the trigeminal ganglion is much more difficult but some speculative methods can be considered.

A potential route of drug delivery that has received little attention is intra-axonal spread, the same route whereby the virus reaches the ganglionic nerve cell body. Injection of the anthracycline antineoplastic drug doxorubicin directly in the nerve of rats resulted in nerve cell body destruction in the corresponding ganglion cells, including those neurons infected with HSV (Iwasaki et al., 1986). Both doxorubicin and the toxic lectin ricin are transported by retrograde axoplasmic transport, producing so-called "suicide transport" (Wiley et al., 1982). Ricin has been used for neuroanatomical study (including transganglionic tracings) of sensory and autonomic nerves (Harper et al., 1980; Yamamoto et al.,1983); and unlike doxorubicin, it can be delivered by intradermal as well as intraneural injection. In a study of latently infected mice inoculated with HSV-1 by the lip route, ricin injection at the same site some two months or more later eliminated HSV-1 ganglionic latency, as assayed by the standard explantation technique (Hino et al., 1988). Bystander neuronal cell death and dense sensory loss would be an unacceptable consequence of ricin therapy, but non-toxic lectins such as wheat germ agglutinin or natural substances such as nerve growth factor also are taken up by sensory nerve terminals and transported by retrograde axoplasmic flow to ganglionic nerve cell bodies (Gonatas et al., 1979; Thoen and Barde, 1980); and these substances could potentially be used as carriers for anti-sense oligonucleotides or other anti-viral agents. We envisage covalent linkage of oligonucleotides to the carrier protein or peptide via a carbon linker arm mRNA. It will also be important to ensure that accessibility to the target is not precluded because of intra-cellular localization of the coupled oligonucleotide carrier protein complex.

The ultimate carrier of anti-sense could be the virus itself. There is intense research in the development of efficient vector systems, based primarily on retroviruses, for delivering genes into somatic cells, especially bone marrow stem cells. Recently, advances in understanding the regulation of herpesvirus gene expression have led to the exploitation of HSV as a vector for introducing genes into the nervous system (Dobson et al., 1990). We envisage using HSV vectors as a logical starting point to evaluate the potential of anti-sense expression of plasmids in controlling HSV infections. As previously noted, stable HSV transcripts termed LATs are present in latently infected neurons at sufficient levels to be detected by Northern blot hybridization (Stevens et al., 1987). Moreover, viruses have been constructed which stably express, via the LAT promotor, non-viral products such as beta-globin RNA and beta-galactosidase in neuronal cells (Dobson et al., 1989; Ho et al., 1989). A live virus "vaccine" deleted in LAT but with anti-sense constructs under the LAT promoter may prove useful for individuals with very frequent recurrences. To be effective, the engineered virus would have to be inoculated in such a way that ganglion cells already infected with wild type virus are colonized with the engineered virus. A most promising HSV vector is a tk-deficient HSV strain, engineered to express anti-sense genes located at both the TK and LAT gene loci. After inoculation, this vector should replicate peripherally in the skin and ascend to the ganglia; but being tk-deficient, this virus should not cause a productive ganglionic infection nor would it be capable of reactivation (Coen et al., 1989). It is anticipated that continuous anti-sense expression via the LAT promoter will block replication of any wild type HSV strain, thus effectively molecularly immunizing the ganglion cell against HSV infection.

ACKNOWLEDGEMENT

This work was supported by EY 08208 and CA 33572 from the National Institutes of Health.

REFERENCES

Buzayan, J.M., Gerlach, W.L., and Bruening, G., 1986, Satellite tobacco ringspot virus RNA: a subset of the RNA sequence is sufficient for autolytic processing, Proc. Natl. Acad. Sci. USA 83:8859.

Chatis, P.A., Miller, C.H., Schrager, L.E., and Crumpacker, C.S., 1989, Successful treatment with foscarnet of an acyclovir-resistant mucocutaneous infection of herpes simplex virus in a patient with acquired immunodeficiency syndrome, N. Engl. J. Med. 320:297.

Cantin, E.M., Puga, A., and Notkins, A.L., 1987, Molecular biology of herpes simplex virus latency, in: "Concepts in Viral Pathogenesis," A.L. Notkins and M.B.A. Oldstone, eds., Springer-Verlag, New York.

Cech, T.R., and Bass, B.A., 1986, Biological catalysis by RNA, Ann Rev. Biochem. 55:599.

Coen, D.M., Kosz-Vnenchak, M., Jacobson, J.G., Leib, D.A., Bogard, C.L., Schaffer, P.A., Tyler, K.L., and Knipe, D.M., 1989, Thymidine kinase-negative herpes simplex virus mutants establish latency in mouse trigeminal ganglia but do not reactivate, Proc. Natl. Acad. Sci. USA 86:4736.

Coleman, J.A., Green, P.J., and Inouye, M., 1984, the use of RNAs complementary to specific mRNAs to regulate the expression of individual bacterial genes, Cell 37:429.

Coleman, J.A., Hirashima, A., Inokuchi, Y., Green, P.J., and Inouye, M., 1985, A novel immune system against bacteriophage infection using complementary RNA (mic RNA), Nature 315:601.

Corey, L., and Spear, P.G., 1986, Infections with herpes simplex viruses, N. Engl. J. Med. 314:686.

Cremer, K.J., Mackett, M. Wohlenberg, C., Notkins, A.L., and Moss, B., 1985, Vaccinia virus recombinant expressing herpes simplex virus type 1 glycoprotein D prevents latent herpes in mice, Science 228:737.

Crumpacker, C.S., 1989, Molecular targets of antiviral therapy, N. Engl. J. Med. 321:163.

DeLuca, N.A., McCarthy, A.M., and Schaffer, P.A., 1985, Isolation and characterization of deletion mutants of herpes simplex virus type 1 in the gene encoding immediate-early regulatory protein ICP4, J. Virol. 56:558.

Dobson, A.T., Sedarati, F., Devi-Rao, G., Flanagan, W.M., Farrell, M.J., Stevens, J.G., Wagner, E.K., and Feldman, L.T., 1989, Identification of the latency associated transcript promoter by expression of rabbit beta-globin mRNA in mouse sensory nerve ganglia latently infected with a recombinant herpes simplex virus, J. Virol. 63:3844.

Ellis, M.N., Keller, P.M., Fyfe, J.A., Martin, J.L., Rooney, J.F., Straus, S.E., Nusinoff-Lehrman, S., and Barry, D.W., 1987, Clinical isolate of herpes simplex virus type 2 that induces a thymidine kinase with altered substrate specificity, Antimicrob. Agents Chemother. 31:1117.

Erlich, K.S., Mills, J., Chatis, P., Mertz, G.J., Busch, D.F., Follansbee, S.E., Grant, R.M., and Crumpacker, C.S., 1989, Acyclovir-resistant herpes simplex virus infections in patients with the acquired immunodeficiency syndrome, N. Engl. J. Med. 320:293.

Forster, A.C., and Symons, R.H., 1987, Self-cleavage of plus and minus RNAs of a viusoid and structural model for the active sites, Cell 49:211.

Galloway, D.A., and McDougall, J.K., 1983, The oncogenic potential of herpes simplex viruses: evidence for a 'hit and run' mechanism, Nature 302:21.

Gontas, N.K., Harper, C., Mizutani, T., Gontas, J., 1979, Superior sensitivity of conjugates of horseradish peroxidase with wheat germ agglutinin for studies of retrograde axonal transport, J. Histochem. Cytochem. 27:728.

Harper, C.G., Gontas, J.O., Mizutani, T., and Gontas, N.K., 1980, Retrograde transport and effects of toxic ricin in the autonomic nervous system, Lab. Invest. 42:396.

Haseloff, J., and Gerlach, W.L., 1988, Simple RNA enzymes with new and highly specific endoribonuclease activities, Nature 334:585.

Hino, M., Sekizawa, T., and Openshaw, H., 1988, Ricin injection eliminates herpes simplex virus in the mouse, J. Infect. Dis. 157:1270.

Hirsch, M.C., and Schooley, R.T., 1989, Resistance to antiviral drugs: the end of innocence (editorial), N. Engl. J. Med. 320:313.

Ho, D.Y., and Mocarski, E.S., 1989, Herpes simplex virus latent RNA (LAT) is not required for latent infection in the mouse, Proc. Natl. Acad. Sci. USA 86:7596.

Hutchins, C.J., Rathjen, P.D., Forster, A.C., and Symons, R.H., 1985, Self cleavage of plus and minus RNA transcripts of avocado sunblotch viroid, Nucl. Acid Res. 14:3627.

Iwasaki, Y., Yamamoto, T., Konno, H., Iizuka, H., and Kudo, H., 1986, Eradication of herpes simplex virus persistence in rat trigeminal ganglia by retrograde axoplasmic transport, J. Virol. 59:242.

Izant, J.G., and Weintraub, H., 1984, Inhibition of thymidine kinase gene expression by antisense RNA: a molecular approach to genetic analysis, Cell 36:1007.

Knipe, D.M., The role of viral and cellular nuclear proteins in herpes simplex virus replication, Adv. Virus Res. 37:85.

Krol, A.R. van der, Lenting, P.E., Veenstra, J., van der Meer, I.M., Koes, R.E., Gerats, A.G.M., Mol, J.N.M., and Stuitje, A.R., 1988a, An antisense chalcone synthetase gene in transgenic plants inhibits flower pigmentation, Nature 333:866.

Krol, A.R. van der, Mol, J.N.M., and Stuitje, A.R., 1988b, Modulation of eukaryotic gene expression by complementary RNA and DNA sequences, BioTechniques 6:958.

Kulka, M., Smith, G.C., Aurelian, L., Fishelevich, R., Meade, K., Miller, P., and Ts's, P.O.P., 1989, Site specificity of the inhibitory effects of oligonucleoside methylphosphonates complementary to the acceptor splice junction of herpes simplex virus type 1 immediate early mRNA 4, Proc. Natl. Acad. Sci. USA 86:6868.

Leib, D.A., Bogard, C.L., Kosz-Vnenchak, M., Hicks, K.A., Coen, D.M., Knipe, D.M., and Schaffer, P.A., 1989a, A deletion mutant of the latency associated transcript of herpes simplex virus type 1 reactivates from the latent state with reduced frequency, J. Virol. 63:2893.

Meignier, B., Jourdier, T.M., Norrild, B., Pereira, L., and Roizman, B., 1987, Immunization of experimental animals with reconstituted glycoprotein mixtures of herpes simplex virus 1 and 2: protection against challenge with virulent virus, J. Infect. Dis. 155:921.

Nusinoff-Lehrman, S., Douglas, J.W., Corey, L., and Barry, D.W., 1986, Recurrent genital herpes and suppressive oral acyclovir therapy: relationship between clinical outcome and in-vitro drug sensitivity, Ann. Intern. Med. 104:786.

O'Hare, P., and Hayward, G.S., 1985, Evidence for a direct role for both the 175,000 and 110,000-molecular weight immediate early proteins of herpes simplex virus in the transactivation of delayed early promoters, J. Virol. 53:751.

Parker, A.C., Craig, J.O., Collins, P., Oliver, N., and Smith, J., 1987, Acyclovir-resistant herpes simplex virus infection due to altered DNA polymerase, Lancet 2:1461.

Paterson, B.M., Roberts, B.E., and Kuff, E.L., 1977, Structural gene indentification and mapping by DNA-mRNA mediated hybrid-arrest cell-free translation, Proc. Natl. Acad. Sci. USA 74:4370.

Puga, A., and Notkins, A.L., 1987, Continued expression of a poly (A+) transcript of herpes simplex virus type 1 in trigeminal ganglia of latently infected mice, J. Virol. 61:1700.

Roizman, B., 1983, "The Herpesviruses 3", Plenum Publishing Corporation, New York.

Roizman, B., and Sears, A.E., 1987, An inquiry into the mechanisms of herpes simplex virus latency, Ann. Rev. Microbiol. 41:543-577.

Smith, C.C., Aurelian, L., Reddy, P.M., Miller, P.S., and Ts'o, P.O.P., 1986, Antiviral effect of an oligo (nucleoside methylphosphonate) complementary to the splice junction of herpes simplex virus type 1 immediate early pre-mRNAs 4 and 5, Proc. Natl. Acad. Sci. USA 83:2787.

Stanberry, L.R., Bernstein, D.I., Burke, R.L., Pachl, C., and Myers, M.G., 1987, Vaccination with recombinant herpes simplex virus glycoproteins: protection against initial and recurrent genital herpes, J. Infect. Dis. 155:914.

Steiner, I., Spivak, J.G., Lirette, L.P., et al., 1989, Herpes simplex virus type 1 latency-associated transcripts are evidently not essential for latent infection, EMBO Journal 8:505.

Stevens, J.G., 1989, Human herpesvirus: a consideration of the latent state, Microb. Rev. 53:318.

Stevens, J.G., Wagner, E.K., Devi-Rao, G.B., et al., 1987, RNA complementary to a herpes virus alpha gene mRNA is prominent in latently infected neurons, Science 235:1056.

Straus, S.E., Takiff, H.E., Seidlin, M., Bachrach, S., Lininger, Di Giovana, J.J., Wester, K.A., Smith, H.A., Lehrman, S.N., Creagh-Kirk, T., and Alling, D.W., 1984, Suppression of frequently recurring genital herpes: a placebo-controlled double-blind trial of oral acyclovir, N. Engl. J. Med. 310:1545.

Spruance, S.L., Stewart, J.C., Rowe, N.H., McKeough, M.B., Wenerstrom, G., and Freeman, D.J., 1990, Treatment of recurrent herpes simplex labialis with oral acyclovir, J. Infect. Dis. 161:185.

Thoenen, H., Barde, Y.A., 1980, Physiology of nerve growth factor, Physiol. Rev. 60:1284.

Ts'o, P.O.P., Miller, P.S., Aurelian, L., Murakami, A., Agris, C., Blake, K.R., Lin, S.B., Lee, B.L., and Smith, C.C., 1987, An approach to chemotherapy based on sequence information and nucleic acid chemistry, in: "Biological Approaches to the Controlled Delivery of Drugs," Annals N.Y. Acad. Sci. 507:220.

Uhlenbeck, O.C., 1987, A small catalytic oligoribonucleotide, Nature 328:596.

Uhlman, E., and Peyman, A., 1990, Antisense oligonucleotides: a new therapeutic principle, Chemical Reviews 90:544.

Willey, D.E., Cantin, E.M., Hill, L.R., Moss, B., Notkins, A.L., and Openshaw, H., 1988, Herpes simplex virus type 1-vaccinia virus recombinant expressing glycoprotein B: protection from acute and latent infection, J. Infect. Dis. 158:1382.

Willey, D.E., Williams, I., Openshaw, H., 1991, Ocular acyclovir delivery by collagen discs: a mouse model to screen anti-viral drugs, Current Eye. Res. , In Press.

Wiley, R.G., Blessing, W.W., and Reis, D.J., 1982, Suicide transport: destruction of neurons by retrograde transport of ricin, abrin, and modeccin, Science 216:889.

Yamamoto, T., Iwasaki, Y., and Konno, H., 1983, Retrograde axoplasmic transport of toxic lectins is useful for transganglionic tracings of the peripheral nerve, Brain Res. 274:325.

Zamencnik, P.C., and Stephenson, M.L., 1978, Inhibition of Rous Sarcoma virus replication and cell transformation by a specific oligodeoxynucleotide, Proc. Natl. Acad. Sci. USA 75:280.

HETEROGENEITY OF A HERPES SIMPLEX VIRUS CLINICAL ISOLATE EXHIBITING RESISTANCE TO ACYCLOVIR AND FOSCARNET

Emanuela Pelosi[1], Karen A. Hicks[1],
Stephen L. Sacks[2], and Donald M. Coen[1]

[1]Department of Biological Chemistry and Molecular Pharmacology
Harvard Medical School
Boston, Massachusetts

[2]Division of Infectious Diseases
University of British Columbia
Vancouver, British Columbia, Canada

SUMMARY

Resistance of herpes simplex virus to acyclovir is a problem of growing clinical importance. Acyclovir-resistance can be due either to mutations in the viral thymidine kinase gene or in the viral DNA polymerase gene. Although clinical resistance has most frequently been associated with thymidine kinase alterations, heterogeneity in clinical isolates has not been addressed frequently. The potential for such heterogeneity has been emphasized by a report describing a pathogenic clinical isolate containing within its population at least one thymidine kinase-proficient DNA polymerase mutant as well as mutants exhibiting thymidine kinase-deficiency (Sacks, et al., 1989). We provide here additional characterization of this isolate and speculations regarding its significance.

INTRODUCTION

Treatment of herpes simplex virus (HSV) infections was revolutionized by the development of acyclovir (ACV). This drug, which consists of guanine linked to an acyclic sugar-like moiety, is converted to ACV-monophosphate almost exclusively by the HSV thymidine kinase (TK) (Fyfe, et al., 1978). ACV-triphosphate, which is produced from the monophosphate by cellular enzymes, is a more potent inhibitor of HSV DNA polymerase (Pol) than of cellular polymerases (Furman, et al., 1979). This doubly-selective mechanism of antiviral action contributes greatly to the high therapeutic index of ACV.

In line with this mechanism, resistance to ACV can arise due to mutations in either the HSV *tk* or *pol* gene. In laboratory settings, *tk* mutations conferring ACV-resistance fall into three classes (Coen, 1991; Larder and Darby, 1984): Mutations

Innovations in Antiviral Development and the Detection of Virus Infection
Edited by T. Block *et al.*, Plenum Press, New York, 1992

that completely abolish TK activity (TK-negative); mutations that decrease activity without abolishing it (TK-partial); and mutations that have little if any effect on thymidine phosphorylation, but substantial effect on ACV phosphorylation (TK-altered). All ACV-resistant *tk* mutants remain sensitive to drugs such as foscarnet (phosphonoformic acid, PFA) that inhibit HSV Pol without requiring phosphorylation by TK. Many, but not all, ACV-resistant *pol* mutants exhibit resistance to PFA to varying degrees.

ACV-resistance is no longer merely a laboratory phenomenon. The first several reports of ACV-resistant HSV isolated from patients in the early 1980's were not convincingly shown to contribute to disease and in some cases ACV treatment did not seem to be impeded (Larder and Darby, 1984). More recently, however, there have been several reports of severe, progressive HSV disease in immunocompromised patients in the face of appropriate ACV therapy (Chatis, et al., 1989; Erlich, et al., 1989; Norris, et al., 1988; Sacks, et al., 1989) with one report indicating a resistance rate of 5% among immunocompromised patients from whom HSV can be isolated (Englund et al., 1990). However, there has been tremendous variation in the severity of disease and there have been few if any virological correlates with clinical outcome.

In this paper, we discuss an ACV-resistant isolate derived from a bone marrow transplant patient who suffered from severe progressive HSV esophagitis despite ACV therapy (Sacks, et al., 1989).

MATERIALS AND METHODS

Cells and Viruses

Vero cells were maintained as previously described (Weller, et al., 1983). Virus isolates 294, 615, 615.3, 615.5 and 615.8 have previously been described (Sacks, et al., 1989). Briefly, isolate 294 was obtained from the mouth of the patient (Vancouver UPN 153) before ACV therapy was commenced (Figure 1). Isolate 615 was obtained from a tongue lesion during a course of ACV therapy while the patient's HSV mucositis and esophagitis were worsening (Figure 1) and was indistinguishable from isolate 294 by restriction endonuclease digestion patterns. Isolates 615.3, 615.5, and 615.8 were derived from individual plaques of isolate 615, plated onto Vero cells under medium containing methylcellulose in the absence of any drug as described (Coen, et al., 1985).

Drug Sensitivity Assays

Plaque reduction assays of drug sensitivities were performed as described using ACV, PFA, and aphidicolin (kindly provided by D. Barry, Burroughs Wellcome Co.; B. Erikkson, Astra Lakemedel; and M. Suffness, National Cancer Institute, respectively) and prepared as described previously (Coen, et al., 1983; Coen, et al., 1985; Coen and Schaffer, 1980).

RESULTS

Clinical Course

Figure 1 depicts the clinical course of the patient from whom isolates 294 (pretreatment) and 615 (treatment) were obtained. This has been summarized previously (Sacks, et al., 1989).

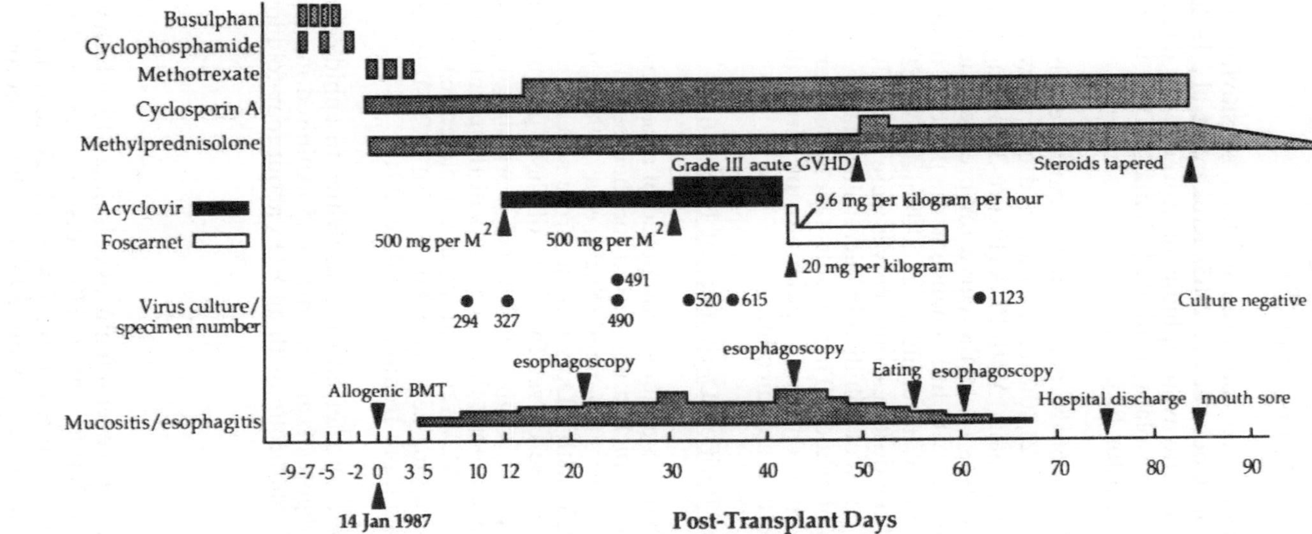

Figure 1. Patient (UPN 153) clinical course following allogeneic bone marrow transplantation. Therapeutic agents, virus culture results, procedures, and clinical status are displayed graphically as a function of time following transplantation. Positive viral cultures are displayed as closed circles next to their respective strain numbers.

Assays of Isolate 615 for Altered Sensitivity to ACV, PFA, and Aphidicolin

Isolate 615, obtained during ACV treatment, was compared with pretreatment isolate 294 for sensitivities to ACV and PFA in a plaque reduction assay (Figures 2 and 3). As previously summarized (Sacks, et al., 1989), 615 was resistant to ACV and PFA, exhibiting doses that reduced plaque formation 50% (ED_{50}'s) roughly fifteen and two-fold higher, respectively, than those of 294.

Isolates 294 and 615 were also compared for their sensitivities to aphidicolin, a drug that inhibits HSV Pol (Coen, et al., 1983; Krokan, et al., 1979), since altered sensitivity, especially hypersensitivity, to aphidicolin is frequently due to *pol* mutations (Coen, et al., 1983). The two isolates exhibited similar ED_{50}'s for aphidicolin; 0.20 μg/ml for 294 and 0.22 μg/ml for 615.

Of interest were the shapes of the dose-response curves exhibited by 615 (Figures 2 and 3). With ACV (Figure 2), the curve flattened out between 20 and 50 μM, before dropping more steeply again. With PFA (Figure 3), a drug that normally gives very steep dose response curves (note the curve for pre-treatment isolate 294), the curve for 615 was relatively flat, with little change in the per cent of plaques remaining between 200 and 400 μg/ml. With aphidicolin, although the ED_{50}'s of 294 and 615 were similar, the shapes of the dose response curves were not. The 615 curve dropped earlier with relatively low doses of aphidicolin decreasing plaque numbers and plaque sizes (Hicks and Coen, unpublished). These results taken together raised the suspicion that 615 might contain different sub-populations of virus that varied in their sensitivities to ACV, PFA, and aphidicolin.

Drug Resistance of Plaque-Purified Derivatives of Isolate 615

To test our suspicion that 615 might be composed of a heterogeneous mixture, ten plaques from isolate 615 were picked and tested for their sensitivities to ACV, PFA, and aphidicolin. As previously summarized (Sacks, et al., 1989), four of these plaques exhibited resistance to ACV with ED_{50}'s ranging from seven- to 30-fold higher than that of 294. The dose-response curves for three of these plaque isolates compared with isolates 294 and 615 are shown in Figure 2, which graphically depicts the variability in ACV-resistance among three members of the population present in 615. Figure 3 shows two of the ACV-resistant plaque isolates, 615.5 and 615.8, were resistant to PFA, again to varying degrees, while a third ACV-resistant isolate, 615.3, was no less sensitive than the pre-treatment isolate, 294.

The plaque isolates were also tested for sensitivities to aphidicolin. Of the three plaque isolates shown in Figures 2 and 3, only 615.5 exhibited more than a marginal alteration in sensitivity to aphidicolin, with an ED_{50} about four-fold lower than that of 294.

DISCUSSION

Sacks et al. (1989) described several features of isolate 615, which was derived from a lesion from a patient suffering severe HSV disease despite ACV therapy. In particular, 615 was found to be resistant to ACV and PFA; to induce HSV TK activity similar to that of the pretreatment isolate, 294; to exhibit neurovirulence similar to that of 294 following intracerebral inoculation of mice; and to be composed of a heterogeneous mixture of different sub-populations, one of which, exemplified by plaque isolate 615.8, contains a *pol* mutation conferring resistance to ACV and PFA. The data presented here graphically depict one aspect of the aforementioned

ACV RESISTANCE OF TREATMENT ISOLATE 615
AND PLAQUE PURIFIED DERIVATIVES

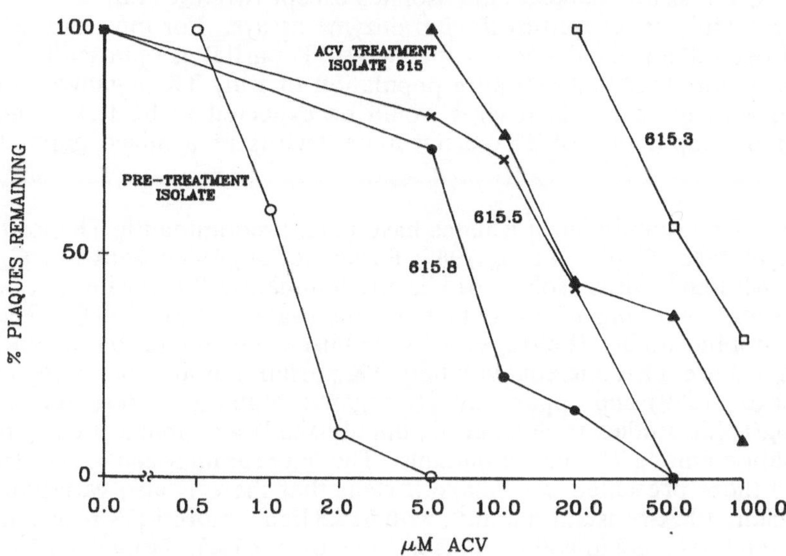

Figure 2. Dose-response of isolates 294, 615, 615.3, 615.5, and 615.8 to ACV in plaque reduction assay.

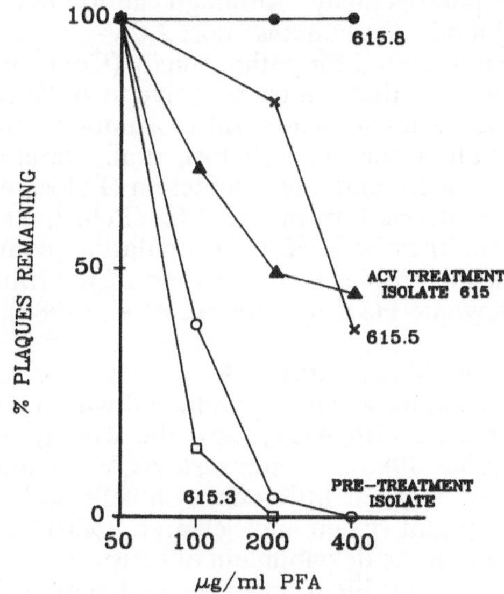

Figure 3. Dose-response of isolates 294, 615, 615.3, 615.5, and 615.8 to PFA in a plaque reduction assay

heterogeneity in terms of differing sensitivities to ACV and PFA and summarize similar results with aphidicolin.

Most ACV-resistant clinical HSV isolates except two, thus far, have exhibited TK-deficiency, mainly as assayed by *in vitro* enzyme assays. For most of these, however, it has not been determined whether they were TK-partial as opposed to TK-negative. Indeed, it is not clear that any pure population of truly TK-negative virus has been associated with disease. Such virus would be expected to be less pathogenic than either a pure population of TK-partial mutant virus or a mixed population (Coen, 1991).

Two ACV-resistant clinical isolates have been predominantly TK-positive, altered DNA polymerase (Collins, et al., 1989; Parker, et al., 1987; Sacks, et al., 1989; this report). Although virus isolate 615 is predominantly TK-positive, it nevertheless contains sub-populations of TK-deficient virus (Sacks, et al., 1989). Recent plaque autoradiographic studies (Pelosi, et al., unpublished) estimate that about 30% of the viruses in 615 are TK-deficient, with both TK-partial mutants such as 615.3 and 615.9 (Sacks, et al., 1989) and apparently TK-negative mutants represented. The plaque autoradiographic studies (Pelosi, et al., unpublished) also make it clear that there is even variation among TK-partial mutants. The data summarized earlier (Sacks, et al., 1989) and those presented here make it clear that there is also variation among the TK-proficient, PFA-resistant mutants, with 615.8 being more PFA-resistant, less ACV-resistant, and less aphidicolin-hypersensitive than 615.5. Finally, 615 appears to contain viruses that are indistinguishable from the pre-treatment isolate in terms of drug sensitivity (Sacks, et al., 1989); we estimate that these viruses represent about 20% of the population (Pelosi, et al., unpublished). Thus, an ACV-resistant isolate derived from a patient failing ACV therapy at the peak of disease is a very heterogeneous mixture of virus.

Why might heterogeneity be important? We speculate that two factors are important. The first is pathogenicity. Although caution must be used in extrapolating from studies in animal models to humans, most ACV-resistant mutants, especially TK-deficient mutants, are attenuated for pathogenicity (Coen, 1991). The assay in mouse models of HSV pathogenesis that is most sensitive to ACV-resistance mutations is the ability of the virus to kill after intracerebral inoculation. Interestingly, isolate 615 is barely attenuated, if at all, in this assay (Pelosi, et al., unpublished; Sacks, et al., 1989) or in a mouse model of acute and latent infection (Pelosi, et al., unpublished), while certain plaque isolates derived from 615 are (Pelosi, et al., unpublished). We speculate that different members of the population complement each other for pathogenicity. This has been observed in assays of neuroinvasiveness with mixtures of otherwise non-pathogenic HSV by Sedarati et al. (1988).

A second factor is ACV-resistance. ACV-resistance due to TK-deficiency is a recessive phenotype; i.e. when a cell is co-infected with a TK-deficient mutant and a wild type virus and treated with ACV, both the wild type and mutant viruses are killed (Coen and Schaffer, 1980). In contrast, ACV-resistance due to altered Pol is co-dominant; ACV treatment has little effect on cells co-infected with wild type and an ACV-resistant pol mutant (Coen and Schaffer, 1980). Thus, *pol* mutations might be particularly important in the development of resistance *in vivo* as mixed populations of virus evolve. Interestingly, Field and Ellis and their colleagues have found that passage of HSV in ACV-treated mice led to production of highly heterogeneous mixtures of virus that were both more resistant and more pathogenic than defined mixtures of TK-deficient or TK-altered and wild type virus (Ellis, et al., 1989; Field, 1982; Field and Lay, 1984). One can speculate that these highly heterogeneous mixtures might resemble isolate 615 and would contain *pol* mutants.

As ACV-resistance in the clinic grows more common, we urge that greater attention be paid to the heterogeneity of resistant isolates and the possibility that such heterogeneity might include *pol* mutations. As data accumulates, it should be possible to correlate these kinds of virological attributes with severity of disease and outcome of treatment.

ACKNOWLEDGEMENTS

We thank D. Barry, B. Erikkson, and M. Suffness for generous provision of the drugs used in plaque reduction assays. This work was supported in part by grants from the NIH (RO1 AI19838, PO1 AI24010, and RO1 AI27209), the Medical Research Council of Canada, and the British Columbia Health Care Research Foundation. E.P. gratefully acknowledges salary support from the University of Verona.

REFERENCES

Chatis, P.A., Miller, C.H., Schrager, L.E., and Crumpacker, C.S., 1989, Successful treatment with foscarnet of an acyclovir-resistant mucocutaneous infection with herpes simples virus in a patient with acquired immunodeficiency syndrome, N. Engl. J. Med. 320:297-300.

Coen, D.M., 1991, The implications of resistance to antiviral agents for herpesvirus drug targets and drug therapy, Antiviral Res., 15:287-300.

Coen, D.M., Furman, P.A., Aschman, D.P., and Schaffer, P.A., 1983, Mutations in the herpes simplex virus DNA polymerase gene conferring hypersensitivity to aphidicolin, Nucleic Acids Res. 11:5287-5297.

Coen, D.M., Fleming, H.E., Jr., Leslie, L.K., and Retondo, M.J., 1985, Sensitivity of arabinosyladenine-resistant mutants of herpes simplex virus to other antiviral drugs and mapping of drug hypersensitivity mutations to the DNA polymerase locus, J. Virol. 53:477-488.

Coen, D.M., and Schaffer, P.A., 1980, Two distinct loci confer resistance to acycloguanosine in herpes simplex virus type 1, Proc. Natl. Acad. Sci. USA 77:2265-2269.

Collins, P., Larder, B.A., Oliver, N.M., Kemp, S., Smith, I.W., and Darby, G., 1989, Characterization of a DNA polymerase mutant of herpes simplex virus from a severely immunocompromised patient receiving acyclovir, J. Gen. Virol. 70:375-382.

Ellis, M.N., Waters, R., Hill, E.L., Lober, D.C., Selleseth, D.W., and Barry, D.W., 1989, Orofacial infection of athymic mice with defined mixtures of acyclovir-susceptible and acyclovir-resistant herpes simplex virus type 1, Antimicrob. Agents Chemother.

Englund, J.A., Zimmerman, M.E., Swierkosz, E.M., Goodman, J.L., Scholl, D.R., and Balfour, H.H., Jr., 1990, Herpes simplex virus resistant to acyclovir: a study in a tertiary care center, Ann. Intern. Med. 112:416-422.

Erlich, K.S., Mills, J., Chatis, P., Mertz, G.J., Busch, D.F., Follansbee, S.E., Grant, R.M., and Crumpacker, C.S., 1989, Acyclovir-resistant herpes simplex virus infections in patients with the acquired immunodeficiency syndrome, N. Engl. J. Med. 320:293-296.

Field, H.J., 1982, Development of clinical resistance to acyclovir in herpes simplex virus-infected mice receiving oral therapy, Antimicrob. Angent Chemother. 21:744-752.

Field, H.J., and Lay, E., 1984, Characterization of latent infections in mice inoculated with herpes simplex virus which is clinically resistant to acyclovir, Antiviral Res. 4:43-52.

Furman, P.A., St. Clair, M.H., Fyfe, J.A., Rideout, J.L., Keller, P.M., and Elion, G.B., 1979, Inhibition of herpes simplex virus induced DNA polymerase activity and viral DNA replication by 9-(2-hydroxyethoxymethyl)guanine and its triphosphate, J. Virol. 32:72-77.

Fyfe, J.A., Keller, P.M., Furman, P.A., Miller, R.L., and Elion, G.B., 1978, Thymidine kinase from herpes simplex virus phosphorylates the new antiviral compound 9-(2-hydroxyethoxymethyl)guanine, J. Biol. Chem. 253:8721-8727.

Hicks, K.A., and Coen, D.M., unpublished results.

Krokan, H., Schaffer, P., and DePamphilis, M., 1979, Involvement of eucaryotic deoxyribonucleic acid polymerases alpha and gamma in replication of cellular and viral deoxyribonucleic acid, Biochemistry 18:4431-4443.

Larder, B.A., and Darby, G., Virus drug resistance: mechanisms and consequences, Antiviral Res. 4:1-42.

Norris, S.A., Kessler, H.A., and Fife, K.H., 1988, Severe progressive herpetic whitlow caused by an acyclovir-resistant virus in a patient with AIDS, J. Infect. Dis. 157:209-210.

Parker, A.C., Craig, J.I., Collins, P., Oliver, N., and Smith, I., 1987, Acyclovir-resistant herpes simplex virus infection due to altered DNA polymerase, Lancet 2:1461.

Pelosi, E., Tyler, K.L., Hwang, C.B.C., and Coen, D.M., manuscript in preparation.

Sacks, S.L., Wanklin, R.J., Reece, D.E., Hicks, K.A., Tyler, K.L., and Coen, D.M., 1989, Progressive esophagitis from acyclovir-resistant herpes simplex: clinical roles for DNA polymerase mutants and viral heterogeneity?, Ann. Intern. Med. 111:893-899.

Sedarati, F., Javier, R.T., and Stevens, J.G., 1988, Pathogenesis of a lethal mixed infection in mice with two nonneuroinvasive herpes simplex virus strains, J. Virol. 62:3037-3039.

Weller, S.K., Aschman, D.P., Sacks, W.R., Coen, D.M., and Schaffer, P.A., 1983, Genetic analysis of temperature-sensitive mutants of HSV-1: the combined use of complementation and physical mapping for cistron assignment, Virology 130:290-305.

CELLULAR METABOLISM AND ENZYMATIC PHOSPHORYLATION OF 9- (2-PHOSPHONYLMETHOXYETHYL) GUANINE (PMEG), A POTENT ANTIVIRAL AGENT

Hsu-Tso Ho, Kathleen L. Woods, Sherry A. Konrad, Hilde De Boeck and Michael J. M. Hitchcock

Bristol Myers-Squibb Pharamaceutical Research Institute
Wallingford, CT

SUMMARY AND INTRODUCTION

PMEG (9-(2-Phosphonylmethoxyethyl)guanine) is a potent, broad spectrum antiviral agent in the nucleoside phosphonate class (Figure 1). *In vitro*, PMEG is active against HSV-1, HSV-2, VZV, HCMV and Rauscher murine leukemia virus with IC_{50}s (the concentrations of drug required to reduce the plaque formation by 50%) less than 1 ug/ml (1,2,3). Similar antiviral activity was demonstrated with TK-minus mutants of HSV-1. However, the toxic effect exerted by PMEG on CEM and Vero cells in culture (TC_{50} 5-10 μg/ml, the concentrations of drug affecting the viability of uninfected cells by 50%) limits the utility of PMEG as an antiviral agent. On the other hand, PMEG has antitumor activity (4) against intraperitoneal P388 leukemia and subcutaneously implanted B16 melanoma in mice. PMEG, also supresses human condylomas from papillomavirus (HPV-11) infected human foreskin in transplanted mice (5). We have investigated the cellular metabolism of PMEG and the viral and cellular target enzymes involved in the mode of action to provide rationale for the low antiviral selectivity and to assist with future design of more selective analogs of PMEG.

RESULTS

Cellular Metabolism and Pharmacology of PMEG

To study the cellular metabolism of PMEG, Vero cells (African Green monkey kidney cells) of 80-90% confluent monolayers were treated with 29 μM or 68 μM of (8-^{14}C)PMEG (sp. act. 22 μCi/mg) for 24 hours. Cells were removed by trypsinization, pelleted and extracted with 60% MeOH/water. Analysis of the cell extract was carried out by anion exchange HPLC using a Whatman Partisil-10 SAX Column, with a potassium phosphate buffer gradient (15-700 mM, pH 3.5). The eluate was then monitored with an on-line Flo-One beta radioactivity detector and a U.V. detector. The elution profile (Figure 2) revealed in addition to PMEG the presence of two products, PMEGp (r.t.= 36.5 min.) and PMEGpp (r.t.= 70.5 min), in the cell extract.

Innovations in Antiviral Development and the Detection of Virus Infection
Edited by T. Block *et al.*, Plenum Press, New York, 1992

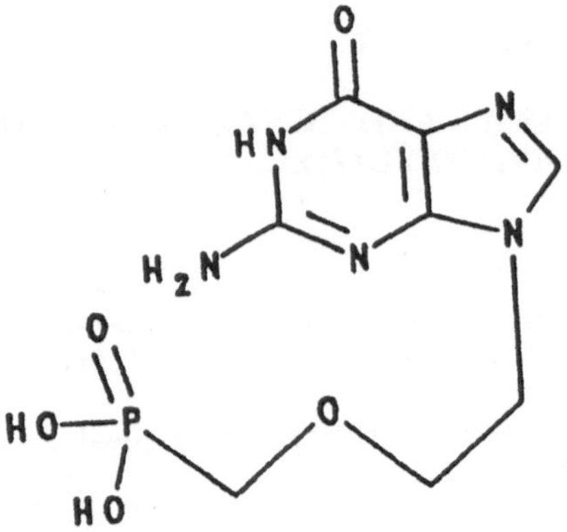

PMEG

Figure 1

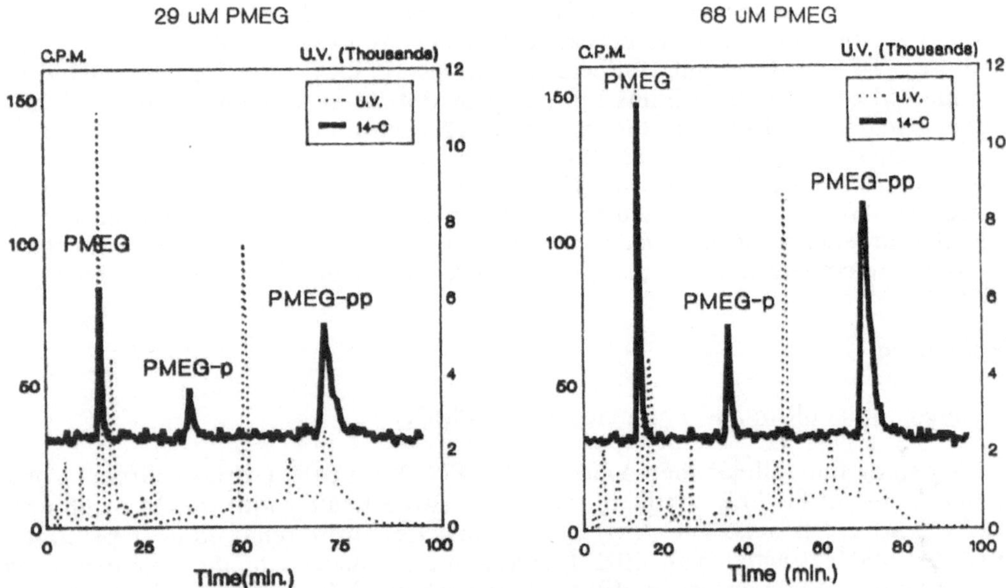

Figure 2. Phosphorylation of 14C-PMEG in Vero Cells.

Table 1. Concentrations of (^{14}C) PMEG and Metabolites in Vero Cells Treated with (^{14}C) PMEG[a]

	Concentration (uM)		
Conc. of (14C) PMEG in media (uM)	PMEG	PMEGp	PMEGpp
29	1.7 (25%)[b]	0.9 (13%)	4.3 (62%)
68	3.4 (26%)	1.9 (14%)	7.9 (60%)

[a] Vero cells were treated with (^{14}C)PMEG for 24 h at 37°C.
[b] Percentage of cell-associated radioactivity.
Cells were grown in 75^2 cm culture flasks.

Table 2. Kinetic Parameters with GMP Kinase and Pyruvate Kinase

	Substrate	Km (uM)	Rel.V_{max}	Rel. (V_{max}/Km), (%)
GMP Kinase[a]	GMP	29.6	100%	100%
	PMEG	226.8	0.9%	0.12%
Pyruvate Kinase[b]	GDP	1.6 mM	100%	100%
	PMEGp	2.4 mM	0.2%	0.13%

[a] Average of 4 experiments
[b] Average of 2 experiments

Table 3. Inhibition of DNA polymerases by PMEGpp

Enzyme	Km (uM) (^{3}H)dGTP	Ki (uM) PMEGpp	Ki (PMEGpp) / Km (dGTP)
HIV-reverse transcriptase	0.29	0.078	0.27
HSV-1 DNA polymerase	0.22	0.006	0.027
DNA polymerase alpha	0.73	0.32	0.44

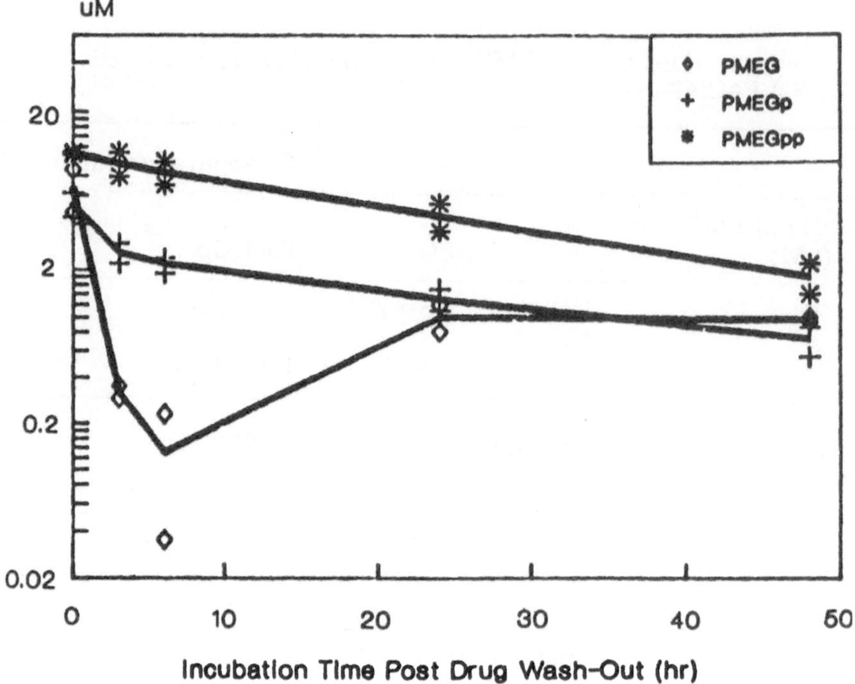

Vero Cells were treated with 44 uM (^{14}C) PMEG for 24 hours before free drug was removed from the media.

Figure 3. Decay of PMEG, PMEGp, and PMEGpp in Vero Cells After Removal of PMEG from the Media.

The identity of the metabolites as PMEG-phosphates was substantiated by their identical retention time with co-injected synthetic standards. The concentration of cell-associated PMEG reached about 5% of that of PMEG in culture media after 24 hours incubation, indicating a very slow rate of cellular uptake. Of the radioactivity associated with the cell extract (Table 1), 60% was attributed to PMEGpp, the presumed active antiviral component.

To evaluate the intracellular persistance of PMEG and its metabolites, Vero cells were first treated with 44 μM ^{14}CPMEG for 24 hours and then incubated further in drug-free media. Cells were processed for analysis at various times after drug-removal. Intracellular PMEGpp decayed with a half-life of 18 hours (Figure 3), as did PMEGp following an initial faster drop. Initially the level of PMEG dropped sharply and reached an equilibrium state after 20 hours.

Enzymatic Phosphorylation of 14CPMEG

To identify the enzymes catalyzing the phosphorylation of PMEG to its mono- and diphosphorylated products, three different enyzme systems were studied with (^{14}C) PMEG as substrate. The formation of ^{14}CPMEGp and/or ^{14}CPMEGpp in these systems was monitored by anion exchange HPLC (see above) of the enzyme reaction mixture after incubation. First,(^{14}C)PMEG was shown to be converted to (^{14}C)PMEGp in a time dependent manner over a 24 hour period when incubated in a reaction mixture containing ATP and porcine brain guanylate kinase (Figure 4a). When the above system was supplemented with either phosphoenolpyruvate and rabbit muscle pyruvate kinase (Figure 4b) or with human erythrocyte nucleoside diphosphate kinase (Figure 4c), the time dependent formation of both (^{14}C)PMEGp and (^{14}C)PMEGpp was evident.

Coupled enzyme assays (6) were set up to determine the phosphorylation kinetic parameters of the target enzymes. Phosphorylation of PMEG or GMP at various concentrations by guanylate kinase (7) was assayed using ATP as the phosphate donor. The formation of nucleoside diphosphates was coupled to the production of pyruvate from PEP (pyruvate kinase) and the reduction of pyruvate to lactate by lactate dehydrogenase was coupled to oxidation of NADH. Reaction rates were measured by the decrease in absorbance at 340 nm with time. The K_m and V_{max} values were derived from a SAS program. Pyruvate kinase catalyzed phosphorylation kinetics of PMEGp or GDP, using PEP as the phosphate donor, was also measured by coupling the reactions to the lactate dehydrogenase catalyzed oxidation of NADH. PMEG was phosphorylated to PMEGp with guanylate kinase at 0.12% efficiency of that of GMP, and PMEGp to PMEGpp with pyruvate kinase at 0.13% efficiency of GDP (Table 2).

Inhibition of Viral and Cellular DNA Polymerases by PMEGpp

The inhibitory effect of PMEGpp on viral and cellular DNA polymerase activities was investigated to evaluate the mechanism of the antiviral action of PMEG. The rate of the polymerase catalyzed incorporation of (^{3}H)dGTP on to the acid insoluble template-primer system (activated salmon sperm DNA) was measured in the presence of various concentrations of PMEGpp and saturating concentrations of dATP, dCTP and TTP. At various times the radioactivity incorporated was counted and the rate of incorporation calculated. The kinetic constants K_m and K_i were determined using a non-linear least square iterative curve fitting based on the equation , $V = (V_{max} \times S)/(S + K_m (1+(I/K_i)))$.

All three DNA polymerases studied were significantly inhibited by PMEGpp (Table 3) in a competitive fashion with respect to dGTP, a natural substrate. HSV-1 DNA polymerase activity was most effectively inhibited by PMEGpp with a $Ki_{(PMEGpp)}/Km_{(dGTP)}$ ratio of 0.027. HIV-reverse transcriptase with $Ki_{(PMEGpp)}/Km_{(dGTP)}$ ratio of 0.27 was ten fold less efficiently inhibited. PMEGpp also inhibited DNA polymerase alpha, a cellular enzyme, with similar efficiency as that of HIV-RT.

DISCUSSION

PMEG, an acyclic guanine nucleoside monophosphate analog with potent antiviral activity *in vitro*, was taken up by and phosphorylated in Vero cells to PMEGp and PMEGpp, with the latter accounting for 60% of the total radioactivity detected in the 60% methanol soluble extract. The concentration of cell associated PMEG was 17 to 20 fold lower than that in the media after 24 hours incubation. Unlike nucleoside analogs such as AZT (8) which rapidly equlibrates through cell membranes by passive diffusion, PMEG uptake is very slow and the mechanism of its cell entry is yet to be determined. Intracellular concentrations reported for HPMPC (9) and PMEA (10) are also low relative to those in the media. Thus, this is probably a general property of nucleoside phosphonates.

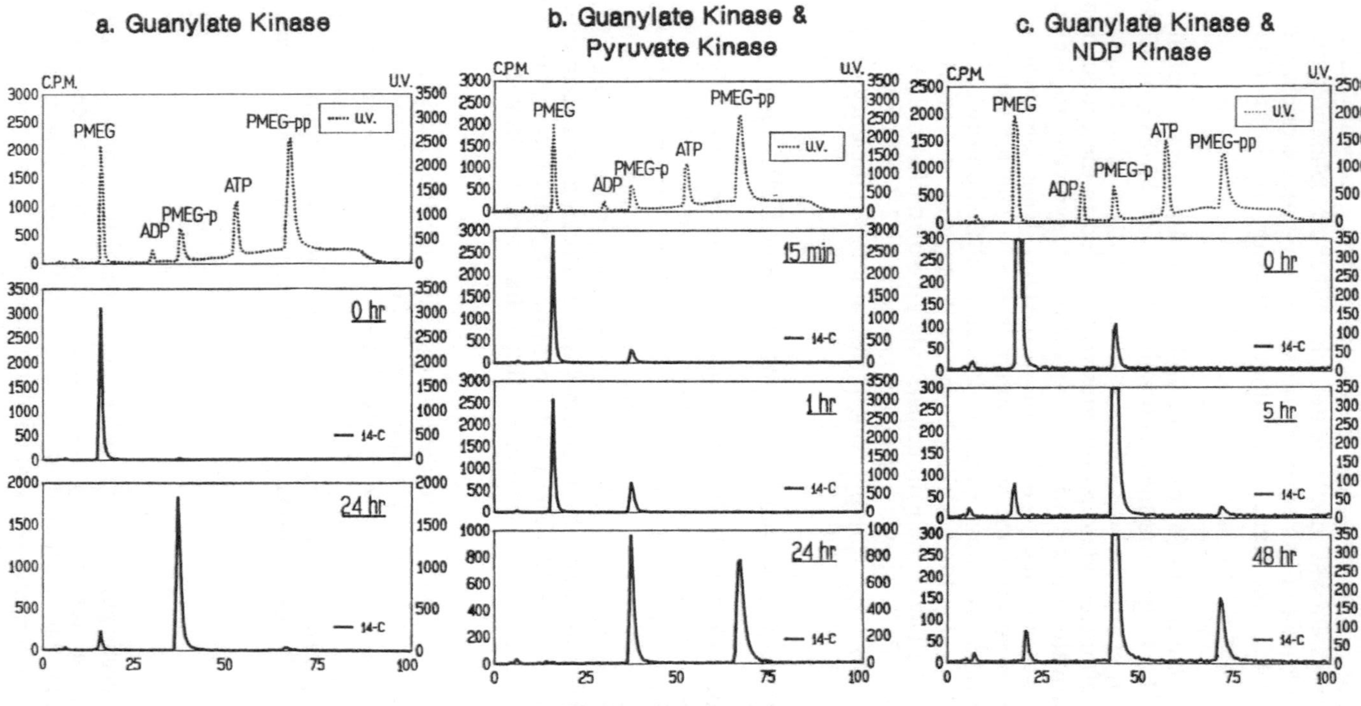

Figure 4. Enzymatic Phosphorylation of 14C-PMEG.

Enzymatic Phosphorylation of PMEG

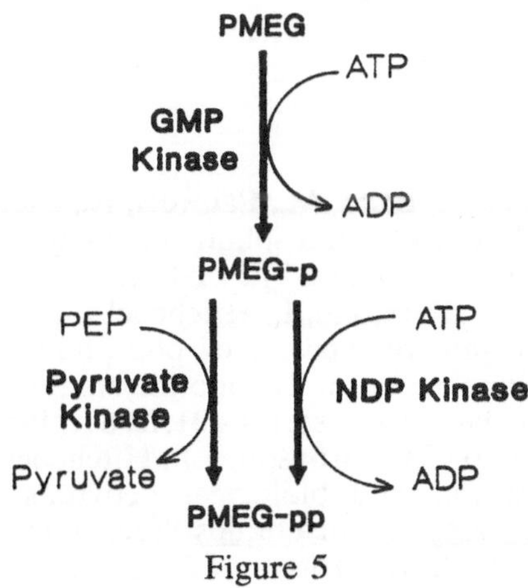

Figure 5

Guanylate kinase was shown to catalyze the phosphorylation of PMEG to its monophosphate, PMEGp, which could be further phosphorylated to PMEGpp by either pyruvate kinase or nucleoside diphosphate kinase. These results suggest that PMEG may be activated intracellularly through the same pathway taken by natural purine nucleotides (Figure 5). In a cell free system, PMEGpp is a very efficient competitive inhibitor of HSV-1 DNA polymerase with respect to dGTP and less so against HIV-reverse transcriptase with the same DNA template-primer. This reflects the *in vitro* antiviral activity demonstrated by PMEG, namely that PMEG is a more potent inhibitor of HSV-replication than HIV-replication in cell culture. Therefore, PMEGpp may be the active component that inhibits the viral replication in cells. Analogs of PMEG with improved *in vitro* selectivity have recently been reported (11). The selectivities of the triphosphate analogs of these compounds with respect to viral and cellular DNA polymerases also reflect the antiviral selectivities shown by the parent drugs in the cell culture (12). These results provide further support that the triphosphate analogs of guanine phosphonates are the active antiviral components. The formation of triphosphate analogs in uninfected cells may therefore render healthy cells resistant to viral replication. Furthermore the phosphorylation of PMEG by cellular enzymes coupled with the low selectivity of PMEGpp between host and viral DNA polymerases provide a rationale for the low selectivity of PMEG as an antiviral agent.

The long intracellular half-life of PMEG-phosphates (18 hours) after drug-removal from the media indicates that the antiviral and anticellular effect of PMEG would persist long after the drug has been cleared from serum. This would suggest that, in animal models of disease, a dosing schedule for PMEG of once a day or less is therefore feasible. It also indicates how the knowledge of the cellular pharmacology can be used to help design appropriate efficacy studies.

ACKNOWLEDGEMENTS

Compounds were supplied by Bristol Myers-Squibb chemists: (^{14}C)PMEG, by Dr. John Swigor; PMEG and PMEGpp, by Dr. Joanne Bronson; PMEGp, by Dr. Kuo-Long Yu.

REFERENCES

1. De Clerq, E., Sakuma, T., Baba, M., Pauwels, R., Balzarini, J., Rosenberg, I., and Holy, A., Antiviral activity of phosphonylmethoxyalkyl derivatives of purine and pyrimidine, Antiviral Res. 8:261-272 (1987).
2. Bronson, J.J., Kim, C., Ghazzouli, I., Hitchcock, J.M., Kern, E.R., and Martin, J., C., Synthesis and antiviral activity of phosphonylmethoxyethyl derivatives of purine and pyrimidine bases, in: "Nucleotide Analogues as Antiviral Agents", J.C. Martin, ed., ACS symposium series 401:72-87 (1989).
3. Kim, C., Luh, B., Misco, P.F., Bronson, J.J., Hitchcock, M.J.M., Ghazzouli, I., and Martin, J.C., Synthesis and biological activities of phosphonylalkyl purine derivatives, Nucleosides & Nucleotides 8(5&6):927-932 (1989).
4. Rose, W.C., Crosswell, A.R., Bronson, J.J., and Martin, J.C., *In vivo* antitumor activity of 9-((2-phosphonylmethoxy)ethyl)-guanine and related phosphonate nucleotide analogues, J. Natl. Can. Inst. 82:50-52 (1990).
5. Krieder, J.W., Balogh, K., Olson, R.O., and Martin, J.C., Treatment of latent rabbit and human papillomavirus infections with 9-(2-Phosphonylmethoxy) ethylguanine (PMEG), Antiviral Res. 14:51-58 (1990).
6. Rudolph, F.B., Baugher, B.W., and Beissner, R.S., Techniques in coupled enzyme assays, Meth. in Enzym. 63:22-42 (1979).
7. Agarwal, K.C., Miech, R.P., and Parks, Jr., R.E., Guanylate kinase from human erythrocytes, hog brain, and rat liver, Meth. in Enzym. LI:483-490 (1978).
8. Zimmerman, T.P., Mahony, W.B., and Prus, K.L., 3'-Azido-3'-deoxythymidine, an unusual nucleoside analogue that permeates the membrane of human erythrocytes and lymphocytes by nonfacilitated diffusion, J. Biol. Chem. 262:5748-5754 (1987).
9. Hitchcock, M.J.M., Woods, K.L., Bronson, J.J., Martin, J.C., and Ho, H.T., Intracellular metabolism of the anti-herpes agent, (S)-1-(3-hydroxy-2-Phosphonyl methoxypropyl) cytosine (HPMPC), 29th Interconference of Antiviral Agents and Chemotherapy, Abstract (1989).
10. Balzarini, J., Hao, Z., Herdewijn, P., Johns, D.G., and De Clerq, E., Intracellular metabolism and mechanism of anti-retrovirus action of 9-(phosphonyl-methoxyethyl) adenine, a potent anti-human immunodeficiency virus compound, Proc. Natl. Acad. Sci. USA 88:1499-1503 (1991).
11. Yang, H., Franco, C., Drain, R.P., and Datema, R., New anti-retroviral acyclic nucleotide analog: (R)-2'-me-PMEG ((R)-N^9-(2-phosphonylmethoxypropyl) guanine), Symposium HIV Disease: Pathogenesis and Therpay, J. Acq. Imm. Def. Syndromes, Abstract, 4:359 (1991).
12. Ho, H.T., De Boeck, H., Woods, K.L., Konrad, S.A., Hitchcock, M.J.M., and Datema, R., Structure and activity correlation-studies of anti-HIV acyclic guanine nucleotide analogs at enzyme level, Symposium HIV Disease: Pathogenesis and Therapy, J. Acq. Imm. Def. Syndromes, Abstract, 4:353 (1991)

EFFECT OF HERPES SIMPLEX VIRUS TYPE 1 INFECTION ON CYTOKINE GENE EXPRESSION IN ACTIVATED MURINE PERITONEAL MACROPHAGES

Linxian Wu[1*], Toby K. Eisenstein[2], Page S. Morahan[1]

[1]Department of Microbiology and Immunology
The Medical College of Pennsylvania
Philadelphia, PA 19129
[2]Department of Microbiology and Immunology
School of Medicine, Temple University
Philadelphia, PA 19140
[*]Corresponding Author

SUMMARY

The intrinsic resistance to herpes simplex virus type 1 (HSV-1) of murine peritoneal macrophages (PMϕ) obtained after *in vivo* infection of different stimuli has been investigated and shown to vary depending on the state of Mϕ activation. Activation of Mϕ by *C. parvum* (CP-Mϕ) or by an avirulent strain of *S. typhimurium* (Sal-Mϕ) increased the permissiveness of Mϕ to HSV-1 infection as evidenced by increased HSV-1 immediate early (IE) gene expression, synthesis of IE proteins, and the degree of cytopathic effect. HSV-1 infection was also found to sharply reduce the level of IL-1-β mRNA in CP-Mϕ) and Sal-Mϕ, and the level of IL-3 mRNA in infected Sal-Mϕ, as measured by northern blot hybridization. Barely detectable levels of IL-β mRNA were found in Sal-Mϕ after infection with HSV-1 when the polymerase chain reaction (PCR) assay was used to confirm the reduction of IL-1-β mRNA. These data suggest that HSV-1 infection can modulate gene expression of some cytokines in the activated Mϕ.

INTRODUCTION

Monocyte-macrophages (Mϕ) are known to be a critical element in antiviral resistance (Morahan et al., 1985; Wu et al., 1990). The intrinsic resistance of Mϕ to virus infection can be affected by the state of differentiation/activation of these cells (Morahan et al., 1985;Wu et al., 1990). The macrophage population resident in tissues is also known to be heterogenous (Morahan et al., 1989a), however, the impact of this heterogeneity on intrinsic resistance to virus infection has not been well characterized. In this investigation, we have compared the intrinsic resistance to HSV-1 infection of murine resident peritoneal macrophages (ResPMϕ) with PMϕ obtained after *in vivo* injection of different stimuli, including thioglycollate broth (TG-Mϕ) (Sit et al., 1988), killed *C. parvum* vaccine (CP-Mϕ) (Sit et al., 1988), and an avirulent strain of *S. typhimurium* (Sal-Mϕ) (Schaffer et al., 1988). Using northern blot hybridization and polymerase chain reaction (PCR) assays, we have also analyzed the effect of HSV-1 infection on cytokine gene expression in these Mϕ.

MATERIALS AND METHODS

Mice

Barrier-raised 6-8 week-old female CD-1 mice (Charles River Laboratories, Inc., Kinston, NY) were received in filter crates and housed in sterile microisolator cages as described previously (Wu et al., 1990a).

Cells and Virus

Vero cells and ResPMϕ were prepared and cultured as described previously (Wu et al., 1990a). Brewers thioglycollate broth (TG) (Difco) was prepared as a 10% solution in distilled water, and aged at least two weeks prior to use as a Mϕ-eliciting agent. Mice were injected intraperitoneally (ip) with 0.5 ml TG, 5 days prior to harvesting peritoneal exudate cells (PEC). The killed vaccine of *C. parvum* was obtained from Burroughs-Wellcome (Research Triangle Park, NC), and injected ip (35 mg/kg) 7 days prior to harvesting PEC. An avirulent strain of *S. typhimurium*, SL3235, was cultured as described (Schaffer et al., 1988). A dose of 1.0-2.6 x 10^6 bacteria was injected ip 7 days prior to harvesting. PEC were harvested from all groups of mice by peritoneal lavage, washed and plated to a 60 mm plastic dish with EMEM. Non-adherent cells were removed by washing three times with PBS after a 2 hour adherence period. HSV-1, strain KOS, was propagated and assayed as described elsewhere (Sit et al., 1988). Cells were routinely infected at multiplicities of 3-5 plaque forming units (PFU) per cell.

Ectoenzyme assays

The assay for alkaline phosphodiesterase activity was carried out on harvested PMϕ as described in detail previously (Morahan et al., 1977).

Immunofluorescence assay

Details of the technique have been described (Sit et al., 1988). Briefly, cells grown on chamber slides (Nunc, Naperville, IL) were fixed in acetone for 15 minutes at 5 hours after infection. Immunofluorescence for HSV-1 antigens was assayed indirectly using a MAb to the HSV-1 IE protein ICP4, followed by fluorescein-labelled rabbit anti-mouse IgG (Sit et al., 1988).

DNA slot blot and northern blot hybridizations

The techniques were performed as we have published (Morahan et al, 1989b; Sit et al., 1988; Wu et al., 1990a). Viral RNA probes including pGEMICP4, pGEMICPO and pGEMTK and their precise construction have been described previously (Morahan et al, 1989b). The murine IL-1-β gene probe was a kind gift of Dr. J. Huang (Sterling Drug, Inc., Malvern, PA) and the murine IL-3 gene probe was purchased form ATCC (Cat. #37552) (Rockville, MD). Several viral DNA probes representing various regions of the HSV-1 genome were used to ensure that the entire genome was present. All plasmids were used as ^{32}P-nick translated probes in either the slot blot or northern blot hybridization experiments, using a kit purchased from Bethesda Research Laboratories (Grand Island, NY).

Polymerase chain reaction (PCR) assay

Isolation of cellular RNA has been described previously (Morahan et al., 1989b; Wu et al., 1991). Reverse transcription was performed on total RNA (1 μg) using 100 U of Moloney murine leukemia virus reverse transcriptase (Bethesda Research Laboratories) and 0.1 μg of primer (dT) (Clontech Laboratories, Inc., Palo Alto, CA) as described (Sambrook et al., 1989). Primers for IL-1-β were purchased from Clontech Laboratories, and PCR was performed as suggested by the supplier. The PCR products were analyzed on a 2% agarose gel and transferred to a nylon membrane. Southern blot hybridization was conducted as described (Sambrook et al., 1989).

RESULTS

Intrinsic resistance to HSV-1 infection in Mϕ stimulated with various activators:

Activaton of murine PMϕ was characterized by a significant decrease in ectozyme activity in Mϕ elicited with any of the three stimuli tested and acquisition of tumoricidal activity by CP-Mϕ and Sal-Mϕ (Eisenstein, et al, 1988; Morahan et al., 1977; Oroskar and Read, 1989; Table 1). All of the activated Mϕ showed increased HSV-1 gene expression as compared with ResPMϕ. TG-Mϕ, CP-Mϕ, and Sal-Mϕ demonstrated various degrees of cytopathic effect (CPE) after HSV-1 infection, in comparison with ResPMϕ which displayed no CPE (Table 1). When infectious virus production was measured, the yield was very low in the ResPMϕ and all of the elicited Mϕs in comparison with Vero cells. However, TG-Mϕ appeared to be slightly more permissive for infectious virus production than the other activated Mϕ (Table 1). The heterogeneity of intrinsic resistance of different elicited macrophage populations was further shown by analysis of HSV-1 DNA replication and viral gene expression in these cells. In regard to production of viral DNA, a progressive decline was found in ResPMϕ and CP-Mϕ, while viral DNA persisted in TG-Mϕ and Sal-Mϕ (Table 1). In regard to viral gene expression, increased levels of mRNA encoding the immediate early gene products ICPO and ICP4 were observed in Sal-Mϕ, in comparison with ResPMϕ and CP-Mϕ. However, mRNA encoding the early gene product, thymidine kinase (TK) was not detected in ResPMϕ, CP-Mϕ or Sal-Mϕ. Using immunofluorescence, a higher percentage of ICP4 positive cells were found in the activated Sal-Mϕ and CP-Mϕ, in contrast to the ResPMϕ (Table 1). Thus, in general, Mϕ activation increased permissiveness to HSV-1 infection, but the degree of permissiveness varied depending on the method of activation.

HSV-1 infection affected IL-1-β and IL-3 gene expression:

Using northern blot hybridization, we analyzed IL-1-β and IL-3 gene expression in Mϕ elicited by different stimuli and infected with HSV-1 *in vitro*. While IL-1-β mRNA was not detected in ResPMϕ, abundant transcripts were detected in mock-infected CP-Mϕ and Sal-Mϕ, which is in agreement with the general concept that activated Mϕ produce high levels of IL-1 (Dinarello 1988). The levels of IL-1-β observed in Sal-Mϕ were approximately 3 fold higher than those seen in CP-Mϕ (Table 2). Following HSV-1 infection, the levels of IL-1-β mRNA were sharply reduced in infected cells (Table 2). IL-3 mRNA was also produced in mock-infected Sal-Mϕ, and was greatly reduced in infected cells (Table 2), while mRNA encoding IL-3 was not detected in mock- or infected CP-Mϕ or ResPMϕ. As a control, the level of ribosomal RNA from these cells was monitored and was not affected at 6 hours post infection by HSV-1 infection (data not shown). It should also be noted that there was no cytopathic effect (CPE) observed in any of the three types of Mϕ at 6 hours post infection, even though severe CPE was observed at 24 hours post infection in both Sal-Mϕ and CP-Mϕ.

In order to determine whether a small amount of IL-1-β mRNA was present in activated Mϕ, but was not detected by northern blot hybridization, we performed PCR.

Table 1. Summary of Various Patterns of Intrinsic Resistance to HSV-1 Infection of Murine Peritoneal Macrophages Elicited with Different Stimuli

Cell Type	Activation			Responses to HSV-1 Infection					
	Ecto-enzyme APD	Tumori-cidal	CPE 24 h pi	Infectious viral (PFU/cell)	[a]Viral DNA Persistence	[b]Viral RNA Synthesis ICP4	ICPO	TK	[c]ICP4 antigen % Positive
Res-PMϕ	4+	-	-	0.003	-	+	-	-	21
TG-Mϕ	+	-	4+	0.340	+	ND[d]	ND	ND	ND
CP-Mϕ	+	+	3+	0.028	-	2+	-	-	73
Sal-Mϕ	+	+	4+	0.020	+	3+	3+	-	75
Vero	ND	ND	4+	191.6	>4+	4+	4+	4+	100

[a] as determined by DNA slot hybridization
[b] Autoradiograms were scanned using scanning densitometry in order to quantitate mRNA
[c] detected by immunofluorescence
[d] not done

Table 2. Summary of Steady-State mRNA Encoding IL-1-β and IL- 3 in Murine Peritoneal Macrophages Elicited with Different Stimuli Using Northern Blot Hybridization

			Optical density reading units		
			hours post infection		
Cell Type	mRNA	0	1	6	24
CP-Mϕ	IL-1β	76.7	46.7	3.1	0.0
	IL-3	0.0	0.0	0.0	0.0
Sal-Mϕ	IL-1β	209.3	227.7	0.0	0.0
	IL-3	17.6	16.1	0.0	0.0

[a] The intensities of the bands from autoradiograms were determined by scanning densitometry.

Abundant IL-1-β transcripts were detected in both mock-infected and infected Sal-Mφ at 1 hour post infection (OD reading units 41.5 and 42.1 respectively). However, IL-1-β mRNA was undetectable in Sal-Mφ at 6 and 24 hours post infection. Thus, the result of the PCR analysis confirmed that IL-1-β mRNA was dramatically reduced in activated Mφ after HSV-1 infection.

DISCUSSION

In this communication, we present additional evidence that activation of Mφ can affect the permissiveness of Mφ to HSV-1 infection (Sit et al., 1988). The results emphasize the fact that heterogeneous patterns of intrinsic resistance exist in Mo obtained using different stimuli. Moreover, this is the first report that demonstrates that HSV-1 infection has a strong impact on cytokine gene expression in Mφ. The reduction of cytokine mRNAs by HSV-1 infection in activated Mφ may be regulated at either the transcriptional or the post-transcriptional level (Wu et al., 1991). A possible mechanism could be degradation of cytokine mRNA by VHS protein, which is known to nonspecifically degrade both viral and cellular mRNAs, with the exception of ribosomal RNA (Krikorian and Read, 1991; Kwong et al., 1988; Oroskar and Read, 1989). Our observation that rRNA was stable in infected CP-Mφ and Sal-Mφ would be consistent with this hypothesis. It is also possible that HSV-1 infection induces cellular factor(s), which nonspecifically or specifically degrade cytokine mRNAs (Malter, 1989). In particular, AUUU sequences frequently occur in 3' or 5' untranslated regions of numerous mRNAs (Shaw and Kamen, 1986), and these are known to potentiate mRNA degradation. Analyzing additional cytokine mRNAs will establish whether the reduction is selective for some cytokines or whether all cytokines are affected, and whether the AUUU sequence is involved in the process. It is unlikely that the observed reduction of mRNA is an *in vitro* artifact, because it has been shown that IL-1-β mRNA can be detected in LPS stimulated murine PMφ after long term culture (Frendl et al., 1990). Finally, HSV-1 infection may shut down the transcription of some cytokine genes in activated Mφ. Nuclear run-on assays should be able to address this question. We are currently approaching these possibilities. It is evident that even limited HSV-1 gene expression can produce profound effects on Mφ functions such as cytokine production. Further investigation of the mechanism has obvious importance for understanding the role of Mφ in immunotherapeutic treatment of herpetic infections.

ACKNOWLEDGEMENTS

This work was supported by NIH grants AI255751 (to PSM), AI24004 (to PSM), AI15613 (to TKE) and National Cancer Center (to LW). We are grateful for the kind gift of IL-1-β gene probe from Dr. J. Huang (Sterling Drug, Inc., Malvern, PA).

REFERENCES

Dinarello, C.A., 1988, Biology of Interleukin 1, FASEB J., 2:108-115.

Eisenstein, T.K., Dala, N., Killer, L., Lee, J-C., and Schaffer, R., Paradoxes of immunity and immunosuppression in Salmonella infection, in: "Host Defenses and Immunomodulation to Intracellular Pathogens", T.K. Eisenstein, W.E. Bullock, and N. Hanna, eds., Plenum Publishing Corp., New York (1988).

Frendl, G., Fenton, M.J., and Beller, D.I., 1990, Regulation of macrophage activation by IL-3. II. IL-3 and lipopolysac-charide act synergistically in the regulation of IL-1 expression, J.Immunol., 144:3400-3410.

Krikorian, C.R., and Read, G.S., 1991, In vitro mRNA degradation system to study the virion host shutoff function of herpes simplex virus, J. Virol., 65:112-122.

Kwong, A.D., Kruper, J.A., and Frenkel, N., 1988, Herpes simplex virus virion host shutoff function, J. Virol., 62:912-921.

Malter, J.S., 1989, Identification of an AUUUA-specific messenger RNA binding protein, Science, 246:664-666.

Morahan, P.S., Glasgow, L.A., Crane, J.L., and Kern, K.R., 1977, Comparison of antiviral and antitumor activity of activated macrophages, Cell. Immunol., 28:404-415.

Morahan, P.S., Connor, J.R., and Leary, K.R., 1985, Viruses and the versatile macrophages, Br. Med. Bull., 41:15-21.

Morahan, P.S., Volkman, A., Melnicoff, M., Dempsey, W.L., , Macrophage hetergeneity, in: "Macrophages and Cancer", G. Heppner and A. Fulton, eds., CRC Press, Boca Raton (1989a).

Morahan, P.S., Mama, S., Anaraki, F., and Leary, K., 1989b, Molecular localization of abortive infection of resident peritoneal macrophages by herpes simplex virus type 1, J. Virol., 63:2300-2307.

Oroskar, A.A., and Read, G.S., 1989, Control of mRNA stability by the virion host shutoff function of herpes simplex virus, J.Virol., 63:1897-1906.

Sambrook, J., Fritsch, E.F., and Maniatis, T., "Molecular Cloning: A Laboratory Manual, Second Edition," Cold Spring Harbor Laboratory Press, New York (1989).

Schaffer, R., Nacy, C.A., and Eisenstein, T.K., 1988, Induction of activated macrophages in C3H/HeJ mice by avirulent Salmonella, J. Immunol., 140:1638-1644.

Shaw, G., and Kamen, R., 1986, A conserved AU sequence form the 3' untranslated region of GM-CSF mRNA mediates selective mRNA degradation, Cell, 46:659-667.

Sit, M.F., Tenney, D.J., Rothstein, J.L., and Morahan, P.S., 1988, Effect of macrophage activation on resistance of mouse peritoneal macrophages to infection with herpes simplex virus types 1 and 2, J. Gen. Virol., 69:1999-2010.

Wu, L-X., Anaraki, F., Morahan, P.S., and Leary, K., 1990a, Transient expression of virus-specific promoters in murine resident peritoneal macrophages, J. Leuk., 48:229-236.

Wu, L-X., Morahan, P.S., and Leary, K., 1990b, Mechanisms of intrinsic macrophage-virus interaction in vitro, Microb. Path., 9:293-301.

Wu, L-X., Morahan, P.S., Leary, K., 1991, Regulation of herpes simplex virus type 1 gene expression in nonpermissive murine resident peritoneal macrophages, Submitted.

PROCEDURE FOR EVALUATION OF NEUTRALIZING ANTIBODY TO CYTOMEGALOVIRUS IN COMMERCIAL INTRAVENOUS GAMMA GLOBULIN PREPARATIONS

Cindy Wordell, Lorena Chambers, Maryann Durante, Sue DiRenzo, Leigh Hopkins, and Donald Jungkind

Departments of Pathology and Pharmacy
Thomas Jefferson University
Philadelphia, PA 19107-4998

INTRODUCTION

Cytomegalovirus (CMV) is a common pathogen complicating the course of renal transplant recipients, as well as the course of bone marrow, cardiac and liver transplant recipients (1-4). In organ transplant patients, CMV is the single most important infection manifesting as either a subclinical infection or a potentially life-threatening disease. The incidence of symptomatic CMV infection in the transplant population is high, 60-90%, compared to immunosuppressed non-transplant patients, 42-66% (2).

Transplant recipients may acquire CMV infection from three sources; the graft itself, blood transfusions or reactivation of a latent infection. Three patterns of CMV infection following transplantation have been described. The first pattern is a primary infection which occurs in those recipients who were seronegative for CMV prior to the transplantation. The source of infection in these patients is the graft itself or blood transfusions from seropositive donors (5-9).

The second source is reinfection with a new strain of CMV, from the donor, in a patient who was seropositive for a different strain prior to transplantation (10). Finally, reactivation of latent virus may occur due to the immunosuppressive agents used in the post-transplant period. Symptomatic disease occurs in those developing primary infection with the donor strain and those who develop a secondary infection. Patients who have reactivation of a latent infection often have mild or asymptomatic disease (5-10).

There is conflicting evidence that the presence of antibody to CMV will be protective against development of disease. There is also evidence that transplant patients with severe overwhelming CMV infection frequently have low titers of circulating antibody during the acute phase of infection and that decreased titers are predictive of a less favorable outcome (6). In seropositive renal transplant recipients, the major risk for developing CMV disease is the number of courses of immunosuppressive therapy given, in particular administration of antilymphocyte globulin. In seronegative recipients, the major risk factor is whether the kidney was harvested from a seropositive donor (8- 9).

Very few studies have been reported in the literature which compare the CMV antibody activity of the currently available polyvalent IGIV preparations (24). In other viral diseases, the presence of neutralizing antibody is thought to be important for the resolution of illness. With several polyvalent IGIV products available, it is important that the CMV antibody content and activity be compared. The purpose of this work was to develop a more rapid neutralization (NT) assay for CMV antibody. Then, using a standard antibody unit (Paul Ehrlich Units), we compared NT activity to that determined by ELISA and agglutination assays.

Four commercial sources of polyvalent IGIV were used to determine the activity of the CMV antibody present in the products, and to determine if CMV titers as measured by commercially available kits correlate with neutralization activity.

MATERIALS AND METHODS

GAMMA GLOBULINS: IGIV for use in this analysis was purchased by the Department of Pharmacy. This was done to minimize bias due to manufacturer preselection of lots for testing. Five lots of Sandoglobulin (Sandoz Pharmaceuticals) and Gammagard (Travenol Laboratories), six lots of Gamimune-N (Cutter Laboratories) and four lots of American Red Cross IGIV were tested. All products were reconstituted with the diluent provided to a final concentration of 5%. Gamimune-N is provided as a 5% concentration by the manufacturer. The investigators measuring the antibody titers were blinded as to the identity of the IGIV being tested.

CMV ELISA: The CYTOMEGELISA enzyme-linked immunoassay was conducted using a commercially available ELISA test (Whittaker Bioproducts, Walkersville, MD) for detection of IgG CMV antibody (25). Testing was done according to manufacturer's recommendations except that diluted gamma globulin was used as sample rather than serum. Gamma globulins were tested in dilutions (1:10, 1:20, 1:40, and 1:80) using CYTOMEGELISA kit diluent. The optical density value selected for calculations of Paul Ehrlich Units was the one nearest the middle of the linear O.D. range. Usually ELISA testing of the 1:40 dilution gave results in the linear part of the slope. Each of the samples were tested in duplicate. Results were repeated to insure reproducibility.

CMV ANTIBODY AGGLUTINATION TESTS: Passive latex agglutination was conducted using CMVScan[R], a commercially available latex agglutination test (26) for detection of both IgG and IgM CMV antibody (BBL Microbiology Systems, Cockeysville, MD). The test was used as described in the manufacturer's recommendations for detection of antibody in serum except that diluted gamma globulin was used instead. The IVIG to be tested was diluted 1:10, 1:100, 1:1000, 1:10,000, and 1:100,000 and tested for agglutination. The titer was tested by a twofold method in duplicate within the final test range. For example: If the 1:1000 titer was positive, but 1:10,000 was negative, then dilutions of 1:1000, 1:2000, 1:4000, and 1:8000 were tested in duplicate for the exact endpoint.

PAUL EHRLICH CMV STANDARD: Lyophilized CMV standard serum (Kindly provided by Dr. Kurth, Paul-Ehrlich Institute, Frankfurt, Germany) was diluted to 10 Paul Ehrlich Units (PE units)/ml. This standard was used in the neutralization, ELISA, and agglutination tests. The value of the PE standard was used to calculate the number of PE units in each of the gamma globulin preparations tested by the three different methods.

CMV GROWTH AND NEUTRALIZATION

Tissue Culture: 1 dram tissue culture vials containing MRC-5 fibroblasts were obtained from ViroMed Laboratories, Mpls., MN. Growth medium was Eagle's MEM with HEPES buffer, 2% fetal calf serum, glutamine,and antibiotics (EMEM).

Neutralization Procedure: CMV ATCC strain AD-169 was used throughout. Stock virus was frozen in EMEM with 15% fetal calf serum at -70° C until ready for use. The titer was monitored and found to be stable during the course of the study. Gamma globulin (GG) solutions were diluted 1:2-1:1024 in EMEM . Sodium bicarbonate was used to neutralize acidic GG solutions. To 0.15 ml of each GG dilution was added 0.15 ml of complement, and 0.3 ml of virus. This was mixed and incubated at 36° C for 1 hour. Final complement concentration was 10 U/ml, and final virus concentration was approximately 2.3-7.0 X 10^2/ml. A 1:10 dilution of the complement did not result in interference through nonspecific neutralization. Final IGIV dilution ranges were 1:8-1:4096. 0.2 ml from each neutralization tube was added to duplicate vials of MRC-5 cells. Vials were centrifuged at 1000 X G for 1 hour at 10° C. The sample was removed, the monolayer was washed twice with buffer, and then fed with EMEM. Incubation was for 6 days before staining. Vials were emptied, washed, fixed, and stained using CMV Direct FA Stain (Bartels Immunodiagnostic Supplies, Inc., Mahwah, NJ). This stain contained antibodies to nuclear and cytoplasmic CMV antigens. The number of micro-plaques/coverslip were counted at 100X and 200X.

RESULTS

A number of possible stains were evaluated before the Bartels stain for CMV microplaques was selected. Dupont anti-immediate early and anti-late antigen stains, plus Serono anti-early, and Whittaker Bioproducts anti-early antigen stains were tested using appropriately shorter incubation times for growth of CMV. These were found to give slightly more inconsistent results for quantitative staining of CMV microplaques.

A typical linearity test using dilutions of our CMV virus stock with no neutralizing antibody present is shown in Figure 1. We found that the counts were linear and countable over a range of approximately ten to over two hundred microplaques. The effect increasing CMV neutralizing activity by increasing the concentration of added IGIV in the microplaque neutralization test is shown in Figure 2. There was a linear reduction in the number of virus microplaques over a wide range as the amount of neutralizing antibody increased.

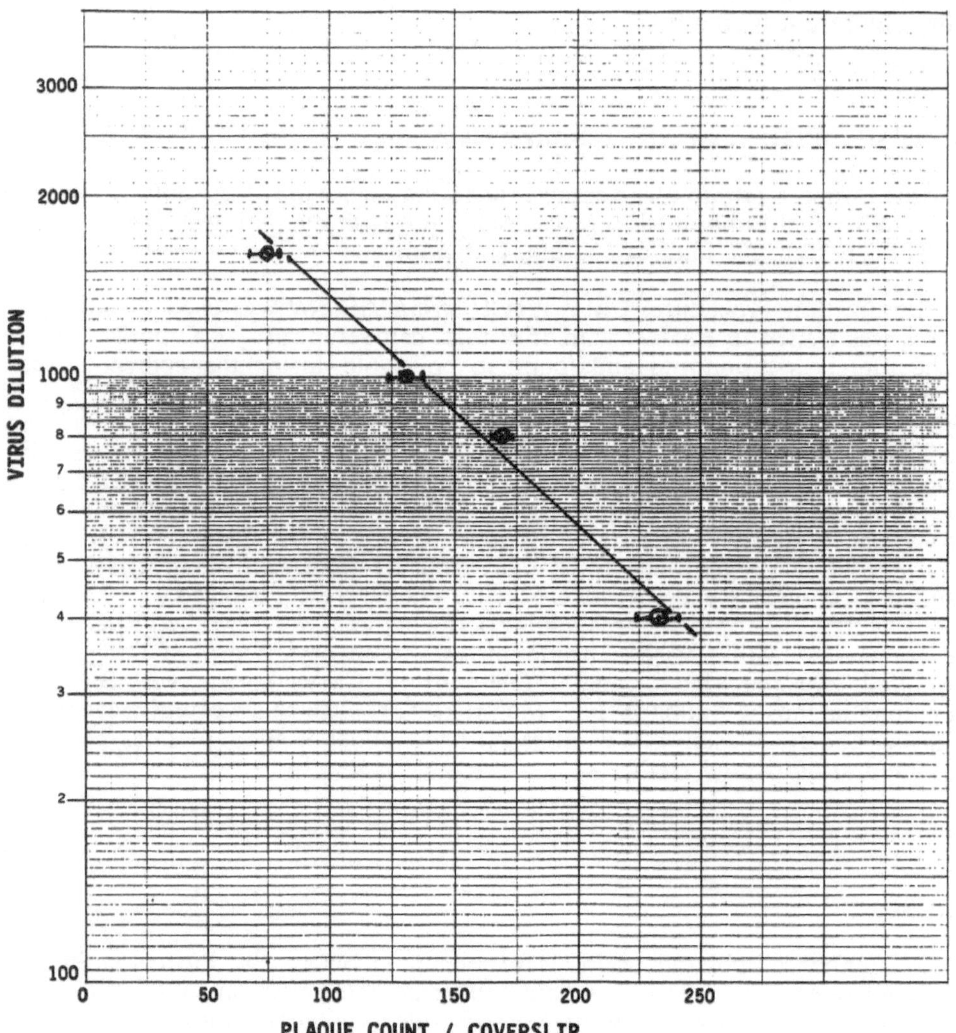

Figure 1. A typical linearity test using dilutions of CMV ATCC strain AD-169 virus stock with no neutralizing antibody present. CMV inoculum was added to duplicate 1 dram vials containing MRC-5 cells. After centrifugation at 1000 X G for 1 hour at 10° C, inoculum was removed and the monolayer was washed twice with buffer, fed with EMEM and incubated for 6 days at 35° C before staining with Bartel's monoclonal antibody CMV specific stain. Microplaques were counted using a fluorescent microscope and 100 X - 200 X magnification.

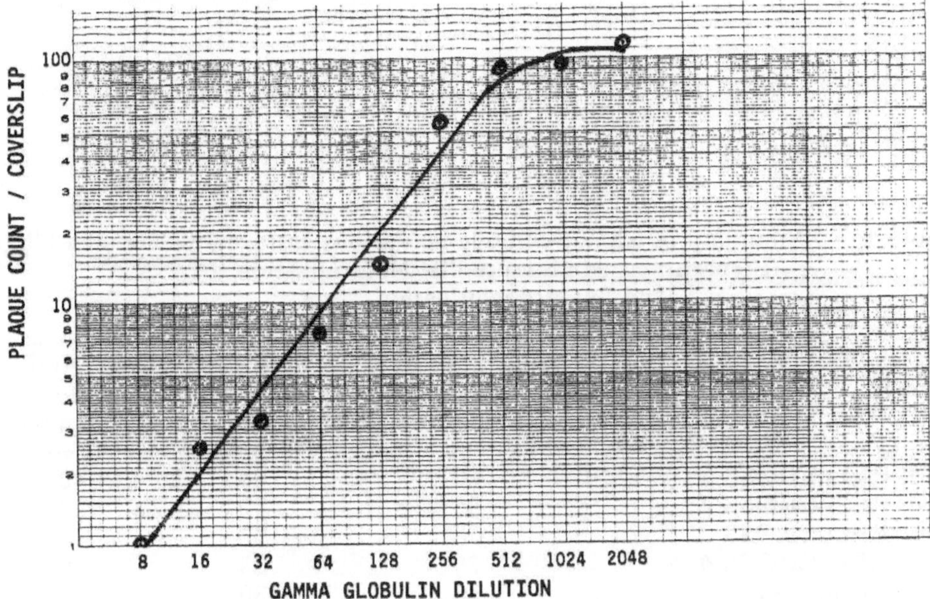

Figure 2. CMV neutralizing activity as a function of IGIV concentration in the neutralization test is shown in this experiment. To 0.3 ml of CMV virus inoculum, 0.15 ml of IGIV and 0.15 ml of complement was added. This was incubated for 1 hour at 36° C. 0.2 ml of this mixture was used to inoculate each of 2 dram vials. Subsequent inoculation and incubation steps were as described in Figure 1.

The CMV antibody activity in 20 lots of gamma globulin from 4 different manufacturers is shown in Table 1. The 80% neutralization endpoint titer of each IGIV preparation was calculated and expressed as PE units after comparison to the 80 % endpoint inhibition titer in the 10 unit PE standard. The 80% neutralization endpoint titer was calculated using the method of proportional interpolation when the endpoint fell between two doubling dilution titers. For comparison, the amount of antibody activity in PE units was also calculated using the CYTOMEGELISA and CMVScan agglutination tests.

In Table 1, Gammagard[R] (Travenol Laboratories) had the highest mean neutralization titer, while Gamimune-N[R] (Cutter Laboratories) had the highest mean ELISA titer. ELISA titer correlated well with neutralization titer for all products except Gammagard[R] which had a mean neutralization titer approximately twice that of the ELISA titer.

Table 2 shows a comparison of the ratio of neutralization titer to ELISA titer. Table 2 shows that PE unit values in the various IGIV lots as determined by 80% neutralization or ELISA were within $\pm$ one twofold difference in 70% of the samples. Ninety-five percent of the samples had values within threefold. This indicated good correlation between ELISA and neutralization titers where $\pm$ one doubling dilution (ratio of 2) is generally considered to be a normal variation that is not significant. A fourfold difference is considered to be the minimum significant difference.

Table 1. Summary of CMV Antibody in Various Intravenous Gamma Globulin Preparations as Measured by Neutralization, ELISA, and Latex Agglutination Tests

Product	Results Expressed As PE Units of CMV Antibody*		
	Neutralization	ELISA	Latex Agglutination
AM. RED CROSS 1	60	24	40
AM. RED CROSS 2	36	40	20
AM. RED CROSS 3	28	32	40
AM. RED CROSS 4	20	23	80
Mean Value 1-4	36	30	45
GAMIMUNE-N 1	107	50	40
GAMIMUNE-N 2	47	30	40
GAMIMUNE-N 3	38	30	20
GAMIMUNE-N 4	30	28	40
GAMIMUNE-N 5	29	50	20
GAMIMUNE-N 6	25	25	80
Mean Value 1-6	46	36	40
GAMMAGARD 1	121	42	20
GAMMAGARD 2	68	28	40
GAMMAGARD 3	51	24	40
GAMMAGARD 4	49	46	40
GAMMAGARD 5	37	26	20
Mean Value 1-5	65	33	32
SANDOGLOBULIN 1	59	15	20
SANDOGLOBULIN 2	25	21	40
SANDOGLOBULIN 3	22	20	40
SANDOGLOBULIN 4	21	18	10
SANDOGLOBULIN 5	15	23	20
Mean Value 1-5	28	19	26
P.E. Std.	10	10	10

* Paul Ehrlich CMV antibody standard test results were used to calculate PE unit values on test results of all other specimens.

Table 2. Comparison of CMV Neutralization Test and ELISA Test Values for 20 Gamma Globulin Preparations

Ratio*	Number Samples	Percent	Cumulative Percent
1.0 - 1.4	11	55 %	55 %
1.5 - 1.9	3	15	70
2.0 - 2.4	3	15	85
2.5 - 2.9	2	10	95
3.0 - 3.4	0	-	-
3.5 - 3.9	1	5	100

* Ratio calculated by dividing PE unit results of the test pair with highest value by the PE units of the test with the lower value. A ratio of 1.0 indicates identical results in both tests.

DISCUSSION

The microplaque neutralization test developed for this study was more convenient and rapid than traditional plaque assay or tube neutralizations for CMV. Bartel's stain which contained two monoclonal antibodies (early and late) made the staining less dependent upon variation in cell culture age, growth rates and antigen expression in the CMV cultures. This stain made it possible to distinguish secondary replication foci from primary microplaques. The use of Paul Ehrlich Standard serum made it possible to more easily cross compare different test methods. Our neutralization test correlated well with the CYTOMEGELISA test in most cases.

Several investigators have demonstrated that both polyvalent and hyperimmune intravenous immunoglobulin (IGIV) preparations containing CMV antibodies positively influence the occurrence of CMV infection and disease in transplant recipients. IGIV has been used for both prophylaxis in those patients at risk of a primary infection and treatment of an acute infection.

Winston et al., demonstrated, in a randomized trial, that CMV seronegative patients undergoing allogeneic bone marrow transplantation experienced fewer episodes of symptomatic CMV disease and interstitial pneumonia than patients not receiving prophylaxis with IGIV (11). A dose of 1 gm/Kg (Gamimune, Cutter Laboratories) was administered prior to transplantation and then weekly for four months. Of those who were seronegative prior to transplantation, 46% of the control group and 21% in the IGIV group had symptomatic CMV disease (p=0.03) while 32% and 16% respectively had CMV interstitial pneumonia (p=0.02). There was also a significant decrease in incidence of graft-versus-host disease (GVHD) in the IGIV group than in the control group.

Tutschka, et al. studied 20 leukemic patients undergoing bone marrow transplantation (12). The patients were stratified into two groups based on their risk for acquiring CMV infection. Seven patients with acute leukemia in remission or chronic leukemia in the chronic phase were in the regular risk group. Thirteen patients with acute leukemia refractory to chemotherapy or in relapse, or with the diagnosis of chronic myelogenous leukemia in accelerated or blastic phase were in the high-risk group. IGIV 0.5 gm/Kg (Sandoglobulin, Sandoz Pharmaceuticals) was administered every other week for 120 days following transplantation. None of the patients developed idiopathic interstitial pneumonia or pneumonia due to CMV. Of the 13 high-risk patients, three died within the first 200 days following transplantation, one from hepatitis B infection and two from adenovirus interstitial pneumonia. The overall survival rate was 85% with all surviving patients in complete remission with functioning grafts.

In a preliminary report, Elfenbein, et al described 82 patients undergoing bone marrow transplantation for hematologic malignancies or aplastic anemia who were enrolled in an open uncontrolled trial evaluating IGIV prophylaxis (13). IGIV (Sandoglobulin, Sandoz Pharmaceuticals) 0.5 gm/Kg was administered prior to the conditioning regimen and then at weekly intervals for 120 days after transplantation. The preliminary evaluation suggested that there were no statistical differences with regard to CMV infection between the two groups.

Most published literature describing IGIV prophylaxis in bone marrow transplantation has used CMV hyperimmune globulin rather than the commercially available polyvalent preparations (14-17). IGIV prophylaxis against CMV following heart transplantation also has used CMV hyperimmune globulin (18). These products have been either produced at the site of the trial or were undergoing clinical studies for a commercial manufacturer. Recently one of the hyperimmune products has received approval for use in prophylaxis of CMV infection following renal transplantation (18).

Like bone marrow transplantation, clinical trials of IGIV in renal transplantation have also primarily used the hyperimmune preparations (19-21). Two reports do described the use of a readily available polyvalent preparation. Steinmuller, et al investigated the use of IGIV for prophylaxis against secondary CMV infection in 34 patients considered to be at high-risk for a secondary infection (22). Sixteen of the patients received IGIV (Sandoglobulin, Sandoz Pharmaceuticals) 0.5 gm/Kg for three doses followed by 0.25 gm/Kg for two additional doses. All doses were administered every two weeks starting at the time the patient was identified as being at high-risk for CMV infection. This had to be within two weeks of transplantation. The number of febrile days and days hospitalized due to CMV disease and the number of complications due to CMV illness were significantly decreased in those who received IGIV.

In contrast to this report, a recent randomized trial found that IGIV in a dose of 0.5 gm/Kg weekly for 12 weeks showed no benefit in prevention of CMV disease in renal transplant recipients (23). This study was conducted in 28 patients receiving cadaver renal transplants. Severe CMV complications did not develop in any patients, whether CMV prophylaxis had been administered or not. In those who did develop CMV disease, IGIV did not improve severity or duration of fever, leukopenia or liver enzyme alterations. The investigators did comment that they have a low baseline incidence of CMV complications in their transplant population. This may have influenced the outcome of this trial. In addition, patients were not preselected for a high-risk of CMV disease, as was done in the study by Steinmuller, et al (22).

In conclusion, the latex agglutination test was not as good for quantitatively determining CMV antibody activity in IGIV preparations as the neutralization and the ELISA tests. The neutralization test detected the widest range of differences in IGIV neutralizing titers. These differences were up to eightfold between the lowest and the highest titers in individual lots. For comparison, a fourfold or greater difference between titers is considered to be significant in most paired serological titer studies. However, no significant difference in anti- CMV titer between manufacturers could be proven using the Newman Keuls test for statistical difference because the number of samples tested was too small. Additional studies need to be performed before final conclusions can be made regarding any differences in anti-CMV activity in polyvalent IGIV preparations and whether these differences are clinically significant.

REFERENCES

Betts, R.F., Cytomegalovirus infection in transplant patients, Prog. Med. Virol., 28:44-64 (1982).

Charpentier, B., Viral infections in renal transplant recipients: An evolutionary problem, Adv. Nephrol., 15:353-378 (1986).

Wreghitt, T.G., Hakim, M., Gray, J.J., Kucia, S., Wallorek, J., English, T.A.H.,Cytomegalovirus infections in heart and lung transplant recipients, J. Clin. Pathol., 41:660-667 (1988).

Wingard, J.R., Yen-Hung, D., Burns, W.H., Fuller, D.J., Braine, H.G., Yeager, A.A.M., Kaiser, H., Burke, P.J., Graham, M.L., Santos, G.W., Saral, R., Cytomegalovirus infection after autologous bone marrow transplantation with comparison to infection after allogeneic bone marrow transplantation, Blood, 71:1432-1437 (1988).

Peterson, P.K., Balfour, H.H., Fryd, D.S., Ferguson, R., Kronenberg, R., Simmons,R.L., Risk factors in the development of cytomegalovirus-related pneumonia in renal transplant recipients, (Letter), J. Infect. Dis., 148:1121 (1983).

Pass, R.F., Griffiths, P.D., August, A.M., Antibody response to cytomegalovirus after renal transplantation: Comparison of patients with primary and recurrent infections, J. Infect. Dis., 147:40-46 (1983).

Glenn, J., Cytomegalovirus infections following renal transplantation, Rev. Infect. Dis., 3:1151-1178(1981).

Smiley, M.L., Wlodaver, C.G., Grossman, R.A., Barker, C.F., Perloff, L.J., Tustin, N.B., Starr, S.E., Plotkin S.A., Friedman, H.M., The role of pretransplant immunity in protections from cytomegalovirus disease following renal transplantation, Transplant., 40:157-161 (1985).

Rubin, R.H., Tolkoff-Rubin, N.E., Oliver, D., Rota, T.R., Hamilton, J., Betts, R.F., Pass, R.F., Hillis, W., Szmuness, W., Farrell, M.L., Hirsch, M.S.Multicenter seroepidemiologic study of the impact of cytomegalovirus infection on renal transplantation, Transplant., 40:243-249 (1985).

Grundy, J.E., Lui, S.F., Super, M., Berry, N.J., Swen,P., Fernando, O.N.,Moorhead, J., Griffiths, P.D., Symptomatic cytomealovirus infection in seropositive kidney recipients: reinfection with donor virus rather than reactivation of recipient virus, Lancet, 2:132-135 (1988).

Winston, D.J., Ho, W.G., Lin, D-H., Bartoni, K., Budinger, M.D., Gale, R.P., Champlin, R.E.,Intravenous immunoglobulin for prevention of cytomegalovirus infection and interstitial pneumonia after bone marrow transplantation, Ann. Intern. Med., 106:12-18 (1987).

Tutschka, P.J., Diminishing morbidity and mortality of bone marrow transplantation, Vox Sang, 51(Suppl 2): 87-94 (1986).

Elfenbein, G., Krischer, J., Graham-Pole, J., Jansen, J., Winton, E., Hong, R., Lazarus, Babington, R., Preliminary results of a multicenter trial to prevent death from cytomegalovirus pneumonia with intravenous immunoglobulin after allogeneic bone marrow transplantation, Transplant. Proc., 14(Suppl 7):138-143 (1987).

Bowden, R.A., Sayers, M., Flournoy, N., Newton, B.,Banaji, M., Thomas, E.D., Meyers, J.D., Cytomegalovirus immunoglobulin and seronegative blood products to prevent primary cytomegalovirus infection after marrow transplantation, N. Engl. J. Med., 314:1006-1010 (1986).

Meyers, J.D., Lewszczynski, J., Zaia, J.A., Flournoy, N, Newton, B., Snydman, D.R., Wright, G.G., Levin, M., Thomas, E.D., Prevention of cytomegalovirus infection by cytomegalovirus immunoglobulin after marrow trans-plantation, Ann. Intern. Med., 98:442-446 (1983).

Winston, D.J., Pollard, R.B., Ho, W.G., Gallagher, J.G., Rasmussen, L.E., Huang, S.N-Y., Lin, C-H., Gossett, T.G., Merigan, T.C., Gale, R.P., Cytomegalovirus immune plasma in bone marrow transplant recipients, Ann. Intern. Med., 97:11-18 (1982).

O'Reilly, R.J., Reich, L., Gold, J., Kirkpatrick, D., Dinsmore, R., Kapoor, N., Condie, R., Randomized trial of intravenous hyperimmunoglobulin for the prevention of cytomegalovirus (CMV) infections following marrow transplantation: preliminary results, Transplant. Proc., 15:1405-1411.

Schafers, H-J., Milbradt, H., Flik, J., Wahlers, T.H., Fieguth, H.G., Haverich, A., Hyperimmunoglobulin for cytomegalovirus prophylaxis following heart transplantation, Clin. Transplant., 2:51-56 (1988).

Snydman, D.R., Werner, B.G., Heinze-Lacey, B., Berardi, V.P., Tilney, N.L., Kirkman, R.L., Milford, E.L., Cho, S.I., Bush, H.L., Levey, A.S., Strom, T.B., Carpenter, C.B., Levey, R.H., Harmon, W.E., Zimmerman, C.E., Shapiro, M.E., Steinman, T., LoGerfo, F., Idelson, B., Schroter, G.P.J., Levin, M.J., McIver, J., Leszczynski, J., Grady, G.F., Use of cytomeagalovirus immunoglobulin to prevent cytomegalovirus disease in renal-transplant recipients, N. Engl. J. Med., 317:1049-1054 (1987).

Greger, B., Valbracht, A., Kurth, J., Schareck, W.D., Muller, G.H., Hopt, U.T., Bockhorn, H., The clinical value of CMV prophylaxis by CMV hyperimmune serum in the kindy transplant patient, Transplant. Proc. 18:1387-1389 (1986).

Fassbinder, E., Ernst, W., Hanke, P., Bechstein, P.B., Scheuermann, E.H., Schoeppe, W., Cytomegalovirus infections after renal transplantation: effect of prophylactic hyperimmunoglobulin, Transplant. Proc., 18:1393-1396 (1986).

Steinmuller, D.R., Graneto, D., Boshkos, C., Novick, A., Cunningham, R., Streem, S., The use of immunoglobulin infusions as prophylaxis for cytomegalovirus infection for living related donor renal transplantation, Am. Soc. Nephrol., 17th Annual Meeting, Washington, D.C., December 13, 1987 (Abstract).

Kasiske, B.L., Heim-Duthopy, K.L., Tortorice, K.L., Ney, A.L., Odland, M., Rao, K.V., Polyvalent immunoglobulin and cytomegalovirus infection after renal transplantation, Arch. Intern. Med., 149:2733-2736 (1989).

Emanuel, D., Cunningham, I., Jules-Elysee, K., Brochstein, J.A., Kernan, N.A., Laver, J., Stover, D., White, D.A., Fels, A., Polsky, B., Castro-Malaspina, H., Peppard, J.R., Bartus, P., Hammerling,U.,O'Reilly, R.J., Cytomegalovirus pneumonia after bone marrow transplantation successfully treated with combination of ganciclovir and high-dose intravenous immunoglobulin, Ann. Intern. Med., 109:777-782 (1988).

Reynolds, D.W., Stagno, S., Alford, C.A., Laboratory diagnosis of cytomegalovirus infections, in: "Diagnostic Procedures for Viral, Rickettsial, and Chlamydial Infections," Fifth Edition, E.H. Lennette and N.J. Schmidt, eds., American Public Health Association, Washington, D.C., (1979).

BBL Microbiology Systems, Division of Becton Dickinson and Company, CMVScan product literature, Cockeysville, MD, April (1986).

IMPROVED DETECTION OF ANTIBODIES TO HEPATITIS C VIRUS USING A SECOND GENERATION ELISA

Stephen Lee[1], John McHutchinson[2], Brian Francis[1], Robert DiNello[3], Alan Polito[3], Stella Quan[3], and Mitchell Nelles[1]

[1]Ortho Diagnostic Systems, Inc.
Raritan, NJ
[2]Division of GI and Liver Disease
University of Southern California
Los Angeles, CA
[3]Chiron Corporation
Emeryville, CA

ABSTRACT

A screening assay for the detection of antibodies to hepatitis C virus (HCV); ORTHO™ HCV ELISA Test System, Second Generation, was compared with the currently licensed c100-3 based test (ORTHO™ HCV ELISA Test System). The second generation ELISA differs from the c100-3 based assay in that it detects circulating antibodies to both structural (nucleocapsid) and non-structural (NS3/NS4) HCV proteins. Specimens tested consisted of a cohort of 35 patients diagnosed with non-A, non-B hepatitis (NANBH) and 3971 presumably healthy volunteer blood donors. Second generation ELISA demonstrated significantly greater clinical sensitivity in patients with acute phase NANBH (80% vs. 60%) as well as chronic disease (88% vs. 72%). Additional specimens reactive only in second generation ELISA, demonstrated reactivity to HCV antigens c33c and/or c22-3 in supplemental testing by the Chiron HCV RIBA ™ Assay System. The second generation ELISA also detected additional RIBA reactive volunteer blood donors (0.18% of the population tested) that were nonreactive in first generation ELISA. This data indicated that second generation ELISA would detect approximately 2 additional anti-HCV reactive donors per 1,000 screened. Specificities obtained with this low risk population were 99.6% for first generation and 99.7% for second generation ELISA.

INTRODUCTION

The successful cloning of non-A, non-B hepatitis (NANBH) agent was first reported in 1989 along with evidence that the agent, now known as hepatitis C virus (HCV), is the major cause of transfusion-associated hepatitis worldwide (Atler et al., 1989; Choo et al., 1989). A cDNA clone derived from the viral RNA was used to produce an HCV recombinant protein, c100-3, which serves as the basis for the first FDA licensed blood screening test (ORTHO™ HCV ELISA Test System) for HCV infection (Agius et. al., 1989; Kuo et al., 1989).

Although the c100-3 based assay has been shown to be an extremely effective methodology for detecting anti-HCV among infectious blood donors and NANBH patients, it was clear from studies early on that delays in seroconversion of up to a year were seen following HCV infection (Alter et al., 1989). In addition, although a large percentage of persons with chronic NANBH developed antibodies to c100-3, a much lower percentage of patients with acute phase disease had detectable levels of anti-HCV. We now report results obtained with a second generation HCV assay which incorporates recombinant proteins derived from 3 distinct regions of the HCV genome. We have assessed the sensitivity of the assay in patients with clinically documented NANBH. In addition, we have evaluated the performance of the test among volunteer blood donors and report on the sensitivity and specificity of the new assay among this low prevalence population.

MATERIALS AND METHODS

Anti-HCV ELISA

Specimens were tested for anti-HCV using the c100-3 based ORTHO™ HCV ELISA Test System and a newly developed ELISA, ORTHO™ HCV ELISA Test System, Second Generation. Both assays are microwell based tests using HCV recombinant antigens. The second generation ELISA differs from the licensed c100-3 based assay, in that it detects antibodies reactive with recombinant proteins derived from 3 distinct regions of the HCV genome. The second generation assay utilizes recombinant antigens derived from the putative NS3 and NS4 regions of the HCV genome which encode the non-structural antigens c100-3 and c33c, and a recombinant antigen c22-3, derived from the putative core (nucleocapsid) region of the HCV genome. For both assays, 20 ul of human serum or plasma were added to 200 ul of specimen diluent in each well of the microtiter plate. After incubation for 1 hour at 37°C, plates were washed 5 times with wash buffer and reacted with 200 ul of mouse monoclonal anti-human IgG conjugated to horseradish peroxidase. Microwells were incubated for 1 hour at 37°C and washed as described above. Following addition of 200 ul of o-phenylenediamine.2HCl substrate solution, microwells were incubated for 30 minutes in the dark at room temperature. Enzyme reaction was terminated by the addition of 50 ul of 4N H_2SO_4 and A_{492} was read. All initially reactive specimens were retested in duplicate. Repeatably reactive specimens were subjected to supplemental testing by Chiron RIBA™ HCV Test System (Second Generation).

RIBA Analysis

Specimens reactive by ELISA were tested using the RIBA™ HCV Test System. This nitrocellulose-based test detects antibodies to the HCV recombinant antigens 5-1-1, c100-3, c33c and c22-3. Testing was carried out as described in the directional insert. Briefly, 20 ul of each specimen was added to 1 ml of specimen diluent and incubated with the immunoblot strip for 4 hours at room temperature. After washing, each strip was then reacted with goat anti-human IgG conjugated with horseradish peroxidase. Following incubation for 30 minutes, strips were washed and incubated with 4-chloro-1-napthol substrate solution for 15 minutes at room temperature. Reaction was stopped by decantation and washing with distilled water. Results were reported as reactive (reactivity with at least two antigen bands), indeterminate (reactivity with a single antigen band) or non-reactive (no reactive bands).

Clinical Specimens

Clinical sensitivity of the new ELISA was assessed using a cohort of 35 serum specimens from patients diagnosed with non-A, non-B hepatitis. The criteria for diagnosis of acute NANB hepatitis include a peak ALT >1000 IU/ml and peak bilirubin >50 μmol/1. Acute specimens were negative for HBsAg, anti-HBc IgM and anti-HAV IgM. Patients were classified as having chronic NANBH when presenting with at least two samples with raised ALT values more than two times normal at least 6 months apart. Serum was not considered to represent chronic NANB hepatitis if the bleed date was within 6 months of an episode of acute hepatitis.

Assay specificity was determined by testing serum and plasma specimens derived from volunteer blood donors. Specimens were from a single geographical site and represented a true random sampling since specimens reactive for other viral markers had been retained.

RESULTS

NANB Hepatitis Specimens

A total of 35 specimens derived from cases of NANB hepatitis were tested by both ELISAs and by RIBA. Of these, 10 were from cases of acute NANBH tested a mean of 36 days after the peak rise in ALT value. The remaining 25 sera were from cases of chronic NANBH of which 12 were asymptomatic with no clinical evidence of liver disease.

The sensitivities of the first and second generation ELISA in acute NANB hepatitis were 60% and 80% respectively, and 72% and 88% in chronic NANBH (Table 1). Twenty-four (69%) specimens were reactive in both the first and second generation ELISA systems. A further 6 specimens (17%) were reactive only in the second generation ELISA and 5 were nonreactive in both ELISAs (Table 2). All specimens reactive with both ELISAs were also reactive when tested by RIBA. All of the specimens that were nonreactive in both ELISAs were nonreactive in RIBA.

Table 1. Reactivity of 1st and 2nd generation ELISA among NANBH patients

	% Patients reactive	
Disease phase	1st generation ELISA	2nd generation ELISA
Acute (N=10)	60%	80%
Chronic (N=25)	72%	88%

Of the 6 specimens reactive only in the second generation ELISA, 3 were reactive in RIBA and 3 were classified as indeterminate (Table 3). Two of these 6 sera were from patients with acute disease and the specimen tested was within 10 days of the peak ALT level. The other 4 specimens were from patients with chronic disease with mildly raised levels of ALT. All of these specimens demonstrated reactivity to c33c and/or c22-3 HCV antigens when tested by RIBA.

Of the 17 serum samples derived from patients with transfusion-associated hepatitis, fourteen were reactive with both first and second generation ELISA, while 3 were reactive only in the second generation ELISA. Similarly, in the nine cases whose principal risk factor was intravenous drug abuse, two were reactive in only the second generation ELISA and seven were reactive in both assays. A further specimen which was reactive only in second generation ELISA, was derived from a patient with chronically raised ALT levels following needle-stick exposure. The 5 sera that were nonreactive in both ELISAs were derived from patients who had no identifiable source of infection or exposure to NANB hepatitis.

Volunteer Blood Donors

A total of 3971 fresh serum specimens from presumably healthy volunteer blood donors were tested with both first and second generation HCV ELISA. The rate of initial reactivity in first generation ELISA was 1.41% with a rate of repeat reactivity of 1.13% (Table 4). Second generation ELISA demonstrated an initial reactive rate of 1.43% with a repeat reactive rate of 1.39%. Ratios of repeat to initial reactivity were therefore 80% for the first generation ELISA and 96% for the second generation ELISA.

In order to evaluate ELISA performance, it was necessary to determine which ELISA reactive specimens also demonstrated the presence of anti-HCV antibodies in supplemental (RIBA) testing. All repeatable reactive specimens from the volunteer donor population were therefore tested further in RIBA (Table 4). Of the 45 specimens repeatably reactive in first generation ELISA, 23 (51%) were reactive when tested by RIBA. In the case of second generation ELISA 30 (55%) were reactive in RIBA. All of the first generation ELISA reactive/RIBA reactive specimens were also reactive in second generation ELISA. Moreover, second generation ELISA detected an additional 7 RIBA reactive specimens, which represented 0.18% of the random population tested. Second generation ELISA also detected an additional 9 specimens that displayed reactivity to a single antigen in RIBA and were classified as indeterminate. Of the specimens repeatably reactive in first generation ELISA, 16 were nonreactive in RIBA, while only 10 samples reactive in second generation ELISA were nonreactive by RIBA. The distribution of absorbances obtained for the volunteer donors was similar in both first and second generation ELISA. The mean absorbance values for ELISA nonreactive specimens was 0.047 +/- 0.046 for first generation ELISA and 0.053 +/- 0.039 for second generation ELISA.

DISCUSSION

The sensitivity and specificity of a newly developed ELISA, ORTHO™ HCV ELISA Test System, Second Generation was investigated in two distinct clinical populations. In this study of 35 NANBH patients, second generation ELISA demonstrated increased sensitivity compared to c100-3 based ELISA, for the detection of anti-HCV antibodies in both acute and chronic disease. The increase in clinical

Table 2. Reactivity of specimens from NANB hepatitis patients by ELISA and RIBA

Disease Phase	ELISA RESULT 1st gen.	2nd gen.	RIBA RESULT 5-1-1	c100-3	c33c	c22-3	Interpretation
Acute	-	-	+/-	+/-	+/-	+/-	Nonreact.
Acute	+	+	3+	+/-	2+	2+	Reactive
Acute	+	+	+/-	2+	2+	4+	Reactive
Acute	-	+	+/-	-	4+	4+	Reactive
Acute	-	+	-	+/-	-	4+	Indeter.
Acute	+	+	4+	4+	4+	4+	Reactive
Acute	-	-	-	-	-	+/-	Nonreact.
Acute	+	+	+/-	4+	+/-	4+	Reactive
Acute	+	+	4+	4+	4+	+/-	Reactive
Acute	+	+	1+	+/-	3+	4+	Reactive
Chronic	-	-	+/-	+/-	+/-	+/-	Nonreact.
Chronic	+	+	4+	4+	4+	4+	Reactive
Chronic	+	+	4+	4+	4+	4+	Reactive
Chronic	-	-	-	-	-	+/-	Nonreact.
Chronic	-	-	-	-	-	+/-	Nonreact.
Chronic	+	+	4+	2+	2+	4+	Reactive
Chronic	+	+	4+	4+	4+	4+	Reactive
Chronic	-	+	+/-	+/-	4+	4+	Reactive
Chronic	+	+	2+	2+	4+	4+	Reactive
Chronic	+	+	-	4+	2+	4+	Reactive
Chronic	+	+	4+	4+	4+	4+	Reactive
Chronic	+	+	4+	4+	4+	4+	Reactive
Chronic	+	+	3+	4+	4+	4+	Reactive
Chronic	+	+	-	2+	2+	4+	Reactive
Chronic	-	+	-	-	4+	4+	Reactive
Chronic	+	+	2+	1+	4+	4+	Reactive
Chronic	-	+	-	+/-	-	4+	Indeter.
Chronic	+	+	4+	4+	4+	4+	Reactive
Chronic	+	+	4+	+/-	4+	4+	Reactive
Chronic	+	+	4+	4+	4+	4+	Reactive
Chronic	+	+	+/-	+/-	4+	4+	Reactive
Chronic	+	+	4+	4+	4+	4+	Reactive
Chronic	+	+	2+	+/-	4+	2+	Reactive
Chronic	+	+	4+	4+	4+	4+	Reactive
Chronic	-	+	-	-	4+	+/-	Indeter.

Table 3. Specimens reactive in second generation ELISA only

Specimen ID#	Diagnosis	ELISA OD 1st. gen.	ELISA OD 2nd gen.	ALT	RIBA reactivity
4	Acute	0.132	2.500	319	c33c,c22-3
5	Acute	0.184	2.500	280	c22-3
18	Chronic	0.347	2.500	48	c33c,c22-3
25	Chronic	0.364	2.500	203	c33c,c22-3
27	Chronic	0.181	2.500	58	c22-3
35	Chronic	0.441	2.500	65	c33c

Table 4. Study of volunteer blood donors using 1st and 2nd generation ELISAs

	ELISA reactivity		RIBA reactivity of ELISA repeat reactives		
	Initial reactive	Repeat reactive	Reactive	Indeterminate	Nonreactive
1st generation ELISA	56 (1.41%)	45 (1.13%)	23	6	16
2nd generation ELISA	57 (1.43%)	55 (1.39%)	30	15	10

Total number of specimens tested = 3971

sensitivity results from the detection of antibodies reactive with additional structural and nonstructural HCV antigen utilized by the second generation ELISA. Moreover, specimens that were reactive in only the second generation ELISA, demonstrated reactivity to HCV antigens when tested by RIBA and these cases presumably represent true HCV infection. It is also noteworthy that the 5 cases in which anti-HCV antibodies were not detected by either ELISA, were not associated with any currently identifiable source of HCV exposure. Although the majority of patients with sporadically acquired NANBH do have detectable anti-HCV antibodies (Hopf et al., 1990; Kuo et al., 1989), some of these individuals may lack an immune response detectable by current assays. Alternatively, there may be another viral (Bradley et al., 1983) or non-viral cause for the liver injury observed in these seronegative patients.

The second generation ELISA also demonstrated improved sensitivity for the detection of anti-HCV antibodies in a low risk population of volunteer blood donors. Previous studies (Contreras and Barbara 1989; Kuhnl et al., 1989; Sircha et al., 1989) using the c100-3 based ELISA, have demonstrated repeat reactive rates in volunteer donors of between 0.5-1.0%. Importantly, the second generation ELISA detected an additional 7 donors (0.18% of the population) that were nonreactive in the currently licensed c100-3 based ELISA, but which were also reactive when tested by RIBA.

Based on these observations, the second generation ELISA can be expected to detect approximately 2 additional anti-HCV reactive donors for every 1000 screened. Also, the prevalence of specimens reactive in second generation ELISA, but nonreactive by RIBA was lower than that observed for the first generation ELISA. When specimens which demonstrated reactivity to HCV antigens in RIBA were excluded from the population, specificity observed was 99.6% for the first generation ELISA and 99.7% for the second generation ELISA. Results obtained in this study, indicate that the ORTHO™ second generation ELISA provides significantly improved serological diagnosis of NANBH. Moreover, the use of this ELISA for blood donor screening, would be expected to further improve the safety of the blood supply through exclusion of HCV infected individuals that are not detectable by current assays.

REFERENCES

Agius, C., Tupper, B., Garcia, G., Botsko, E., Malka, E., Le, A.V., Krauledat, P., DiNello, R., Quan, S., Chien, D., Polito, A., and Nelles, M., 1989, ORTHO HCV Antibody ELISA Test System for the detection of antibodies to hepatitis C (non-A, non-B hepatitis) virus, (Abstr.), 1st International Meeting on Hepatitis C Virus, Rome, Italy.

Alter, H.J., Purcell, R.H., Shih, J.W., Melpolder, J.C., Houghton, M., Choo, Q.-L., and Kuo, G., 1989, Detection of antibodies to hepatitis C virus in prospectively followed transfusion recipients with acute and chronic non-A, non-B hepatitis, N. Engl. J. Med. 321:1494-1500.

Bradley, D.W., Maynard, J.E., Popper, H., Cook, E.H., Ebert, J.W., McCaustland, K.A., Schable, C.A., and Fields, H.A., 1983, Post transfusion non-A, non-B hepatitis: physiochemical properties of two distinct agents J. Infect. Dis. 148:254-265.

Choo, Q.-L., Kuo, G., Deiner, A.J., Overby, L.R., Bradley, D.W., and Houghton, M., 1989, Isolation of cDNA clone derived from a blood borne non-A, non-B viral hepatitis genome, Science 244:359-362.

Contreras, M., and Barbara, J.A.J., 1989, Screening for hepatitis C antibody, Lancet ii:796-797.

Hopf, U., Moller, B., Kuther, D., Stemerowicz, R., Lobeck, H., Ludtke-Handjery, A., Walter, E., Blum, H., Rogegendorf, M., and Deinhardt, F., 1990, Long-term follow-up of post-transfusion and sporadic chronic hepatitis non-A, non-B and frequency of circulating antibodies to hepatitis C virus (HCV), J. Hepatol. 10 (1):69-76.

Kuhnl, P., Seidl, S., Stangel, W., Beyer, J., Sibrowski, W., and Flik, J., 1989, Antibody to hepatitis C in German blood donors, Lancet ii:324-325.

Kuo, G., Choo, Q.-L., Alter, H.J., Gitnick, G.L., Redeker, A.G., Purcell, R.H., Miyamura, T., Dienstag, J.L., Alter, M.J., Stevens, C.E., Tegtmeier, G.E., Bonino, F., Colombo, M., Lee, W.-S., Kuo, C., Berger, K., Shuter, J.R., Overby, L.R., Bradley, D.W., and Houghton, M., 1989, An assay for circulating antibodies to a major etiologic virus of human non-A, non-B hepatitis, Science 244:362-364.

Sircha, G., Bellobuono, A., Giovanetti, A., and Marconi, M., 1989, Antibodies to hepatitis C virus in Italian blood donors, Lancet ii:797.

DIAGNOSTIC VIROLOGY - THEN AND NOW

Thomas F. Smith, Arlo D. Wold, and Mark J. Epsy

Section of Clinical Microbiology
Mayo Clinic and Mayo Foundation
Rochester, MN 55905

INTRODUCTION

Several technical approaches are required for the laboratory diagnosis of viral infections. For example, over a 30-year span, laboratory methods have ranged from serology for demonstrating acute phase infection or for measuring immunity or evidence of past infection with a virus and use of cell cultures for demonstration of cytopathic effects (CPE) to amplification and detection of nucleic acid sequences of agents by the polymerase chain reaction (PCR). Between these extremes, enzyme immunoassay (EIA), immunologic detection of early antigens, and nucleic acid probes have expanded our ability to diagnose viral infections much more rapidly than in the past. Because the types and prevalence of viral infections vary according to the specific clinical practice, i,e, children's hospital, tertiary care clinic, or State Public Health Laboratory, each individual virology laboratory must design their service tests to support these special needs. In a children's hospital, methods for the diagnosis of respiratory tract are of greatest significance, whereas detection of herpesvirus infections is most important in tertiary care practice. Further, epidemiologic goals of State Health Laboratories may indicate use of serology for enterovirus serotyping or conventional tube cell cultures exclusively for detection of influenza virus CPE.

Technologic developments such as monoclonal antibodies and genetic probes. many of which are marketed in kit formats, have broadened the scope of most laboratories to be able to detect medically important viruses with greater sensitivity and specificity than in the past, Importantly, the speed and efficiency of diagnostic virology has been vastly improved within the past years and has been recognized for significantly adding to the immediate medical management and specific treatment of patients.

In this communication, we review the application of laboratory methods that have been used over the years in diagnostic virology and suggest new dimensions in these effort through use of PCR technology.

Serology

IgG class antibodies can be measured for evaluation of immunity as for rubella virus, measles virus and varicella-zoster virus or the diagnosis of infection such as human immunodeficiency virus (HIV) or hepatitis C virus (HCV). Importantly, the

Innovations in Antiviral Development and the Detection of Virus Infection
Edited by T. Block *et al.*, Plenum Press, New York, 1992

routine screening for antibodies to CMV is performed with sera from organ transplant patients since the presence of antibodies to this virus is a risk factor for symptomatic disease due to this virus infection in a seronegative recipient.

Conversely, acute viral infections such as caused by EBV, rubella, and measles virus can be recognized by the demonstration of specific IgM class antibodies to the virus. Thus, the clinical utility of serology in modern diagnostic virology is generally for the assessment of immune status to viruses (especially those for which reactivated infections have not been recognized), or the recognition of specific IgM class antibodies to an agent which is difficult to detect by cell cultures or by immunologic methods.

Only a few viruses (influenza virus, enterovirus, and herpes simplex virus) are routinely capable of being detected rapidly (24-48 h post inoculation) in cell cultures by the recognition of specific cytopathic effects (CPE). Importantly, the maximum sensitivity of the conventional tube cell culture system is not achieved for several days after the culture is inoculated with the specimen. The main disadvantage of cell cultures is the inability of these systems to detect important viruses such as Epstein-Barr virus, Norwalk and rotaviruses, hepatitis viruses, papillomavirus, parvovirus, and many others. Because of this limitation of cell cultures, clinical laboratories have sought new technologies to provide expanded diagnostic results.

Enzyme Immunoassay

The principle of EIA tests is like that of fluorescent antibody methods, except that the endpoint of the assay is a color change that can be objectively measured spectrophotometrically. The EIA procedure is most often used for detection of viral antigens that are captured by homologous antibodies bound to a solid phase such as a microtiter plate or a bead.

The detection of viral antigens directly in specimen material by EIA generally lacks sensitivity compared with recovery of viruses in cell culture. For example, detection of HSV directly in specimen extracts has yielded sensitivities of only 64 to 77 % compared with cell culture isolation. On the other hand, amplification of the virus in cell culture for a short time (24-48 h), followed by EIA for viral antigens, provided detection of 98% of the isolates eventually recovered in cell culture (Michalski et al., 1986).

Modifications of EIA detection systems using biotin-strepavidin have produced results of sensitivity (90% - 100%) and specificity (99% - 100%) generally equivalent to conventional cell cultures which require several days for maximum results. (Table).

Recently, single-test-unit membrane EIA kits have become available for the rapid ~10 min) detection of viral antigens directly from patient specimens. In one comprehensive evaluation, the Abbott TESTPACK RSV assay was compared with culture and direct immunofluorescence of nasaopharyngeal cells for detection of the virus (Swierkosz et al., 1989). Of 234 extracts of nasopharyngeal cells, 70 (30%) were culture positive, 103 (44%) were positive by direct immunofluorescence, and 112 (48%) were TESTPACK RSV positive,. Overall, of 122 specimens positive by culture or immunofluorescence, TESTPACK RSV detected 108 specimens (sensitivity, 89%).

In another study, the Kodak SureCell Herpes test kit (Eastman Kodak Co., Rochester, NY), a 15-min assay, detected HSV with sensitivity of 100% (vesicular lesions) to 76% for nonvesicular lesions (Dorian et al., 1990). Thus, specimens which contain high-titered virus correlated well with cell culture results.

Table 1. Detection of Herpes Simplex Virus by Herpchek

Reference	No. Specimens	Sensitivity, % Cell Culture	Herpchek	Specificity, %
Langston, et al.	28	27 (96)	28 (100)	100
Dascal, et al. (a)	119	119 (100)	113 (95)	100
Baker, et al.	79	79 (100)	77 (97)	100
Dascal, et al. (b)	68	62 (91)	66 (97)	
Wu, et al.	49	49 (1000	44 (90)	99
Dunkel, et al.	30	27 (90)	30 (100)	100

Another system, Directigen FLU-A, was compared prospectively to isolation in cell culture and direct immunofluorescence for the detection of influenza virus type A (Waner et al., 1991). Directigen FLU-A compared favorably in sensitivity (100%) and specificity (91.6%) with isolation in cell cultures, however, there were 14 false-positive Directigen results, 11 of which gave weak positive reactions. The authors recommend that the assay would be useful in laboratories as an addition to or substitute for direct immunofluorescence tests.

Other similar membrane EIA assays are currently available and certainly more will be developed. Importantly, it must be noted that several individual viruses may cause a particular syndrome; therefore, multiple membrane EIA tests must be used (expensive) or cell cultures must be inoculated as a back-up system to provide the widest possible screen for viral agents.

Immunofluorescence

Direct detection of viral infection cells. Detection of viral antigens in cells collected by scraping affected dermal areas or obtaining throat washings can provide rapid results. Success of this technique is highly dependent on optimal specimen collection, interpretation of results by a highly skilled technologist, and high quality fluorescence equipment. In one study, direct detection of CMV-infected cells (urine and bronchoalveolar lavage) with a monoclonal antibody to the virus was only 50% to 75% as sensitive as specimens inoculated into shell vials (Lucas et al., 1989). Best results have ben achieved for the detection of respiratory syncytial virus (RSV) and varicella-zoster virus (VZV). Accordingly, it is important to recognize that these viruses (RSV, VZV) do not replicate in laboratory cell cultures optimally, and therefore, direct detection comparisons may be falsely high.

Amplification of viruses in cell culture prior to immunofluorescence - shell vial assay. An alternative to the detection of antigens in cells or tissues obtained directly from the patient is the amplification of the virus in cell cultures for a short period of time with subsequent immunofluorescence staining of the infected monolayers for detection of early viral antigens. By this procedure, a low titer of virus present in a specimen can replicate in cell cultures for a short time, thereby amplifying the virus and the number of infected cells compared to the original specimen. For the herpesvirus in particular, this increase in the quantity of antigen is generally reflected by a greater sensitivity of a fluorescence assay compared to the direct detection of viral infected cells obtained from the patient. In addition, debris, consisting of broken cellular frgaments and other background material, in a specimen extract or scraping can be detrimental to the specific interpretation of the fluorescence reaction and is substantially reduced after inoculation into shell vial cell cultures (Gleaves et al., 1984; Minnich et al., 1988, Shuster et al., 1985).

Several variables must be controlled to provide the most sensitive and rapid results by the shell vial assay:

1. Cell Cultures

   Use confluent monolayer cell cultures 3-8 days after preparation (Fdorko et all., 1989). Pretreatment of cell monolayer with dimethylsulfoxide and dexamethasone does produce increased infectivity of cytomegalovirus in some laboratories (Thiele and Woods, 1988; West et al., 1988), but not in others (Epsy et al., 1988). At least three shell vials should be inoculated with blood specimens; two shell vials are probably adequate for inoculation with urine, and bronchoalveolar specimens.

2. Specimens

   Inocula of 0.3 to 0.5 ml produce more fluorescent foci than lower volumes; however, the background debris can cause nonspecific fluorescence and higher volumes of specimen material can cause toxicity producing morphologic changes in host cells. An inoculum of 0.2 ml provides a reasonable balance between optimal sensitivity of the assay and interfering background of debris that can be a problem with the shell vial assay when higher volumes are used. The pH of urine specimens need not be adjusted to neutrality (Gray et al., 1989). Sonicvation of specimens prior to inoculation has improved infectivity of CMV in shell vials in two studies (Agha et al., 1988; Wunderli et al., 1989).

3. Centrifugation

   Low-speed centrifugation (700 x g) is necessary and adequate. Higher speeds can adversely affect cell monolayers (Hudson, 1988). The centrifugation step is probably the most significant technical consideration for obtaining optimal results by the shell vial assay.

4. Incubation

   Sixteen hours at 36° is adequate if other variables are optimal for detection of CMV and HSV. Incubation of shell vials inoculated with HSV for 8 h reduced the sensitivity by about one-third compared to incubation for 20 h (Tac et al., 1989). Detection of HSV (16 h to 20 h) after inoculation in shell vials is generally (90-100%) comparable in sensitivity to isolation of the virus in conventional tube cell cultures (Gleaves et al., 1985; Peterson et al., 1988; Ziegler et al., 1988).

5. Staining of coverslips

   After the cells have been fixed with acetone, it is important that the cover slips are allowed to dry completely. Residual acetone can produce trapping of the monoclonal antibody and therefore cause nonspecific staining.

The indirect test is better than the FITC-labeled monoclonal antibody used in the direct procedure (Jespersen et al., 1989). The choice of fluoresceinated anti-mouse conjugate seems to be relatively unimportant regarding specificity for the H or L chain, Fc, or $F(ab)^2$ components of the IgG antibody molecule.

6. Examination of cover slips

Use high quality equipment capable of detecting fluorescent foci as determined by examination of cells with a laboratory strain of the virus.

Nucleic Acid Detection Methods

Probe Nucleic acid probes will likely supplement rather than replace many existing rapid viral detection mehtods. Special applications may include *in situ* hybridization of viral infected cells or tissues that have been treated with formalin and embedded in paraffin (Hsia et al., 1989; Jiwa et al., 1989; McClintock et al., 1989; Sorbello et al., 1988; Stockl et al., 1989; Van der Bij et al., 1988; Van der Bij et al., 1989). Nucleic acid probes will be particularly valuable for the detection of many viruses that are not routinely recovered in cell cultures.

Clinically, we evaluated immunostaining (monoclonal and polyclonal antibodies) with *in situ* DNA hybridization for the detection of CMV in 108 liver biopsy specimens from 20 liver transplant patients with CMV hepatitis as defined by histological criteria (Paya et al., 1990). In 5 of 20 patients, the diagnosis of cytomegalovirus hepatitis could have been made earlier (mean = 9.6 d) by immunostaining with a monoclonal antibody. Of 47 biopsy specimens with CMV inclusion bodies, the sensitivity and specificity of the diagnostic procedures were immunostaining with monoclonal antibody (72% and 100%). Thus, in our experience, *in situ* DNA hybridization provided a complementray technique, but could not replace immunostaining procedures.

Polymersase chain reaction. PCR technology will add a new dimension to diagnostic virology. For example, HSV has been rarely recovered from CSF samples of patients with encephalitis; however, two recent reports indicated the potential for this technology to detect viral sequences from this source (Aurelius et al., 1991; Rowley et ali., 1990). Other viruses, such as CMV, polyomavirus (JC), EBV, and HIV represent other obvious etiologic agents of CNS disease for which alternative diagnostic techniques are lacking.

Uncomplicated EBV infections can be diagnosed in the laboratory by serologic tests; however, invasive lymphoproliferative processes in the setting of immunosuppression present management problems for the physician. For example, we examined the DNA for EBV sequences using PCR from tissue specimens (nodes, bone marrow, liver, brain, lung, pleura, parotid gland, and skin) of 27 patients with lymphoproliferative disorders and leukemias. EBV sequences were present in tissue fromall 10 immunocompromised patients with lymphoproliferative disorders but only in 1 of 19 tissues from patients with other malignant B- or T-cell lymphomas or leukemias (Telenti et al., 1990).

The clinical utility of PCR in the laboratory diagnosis of EBV infections is demonstrated by the following case. A 25-year old female received a liver transplant for chronic active hepatitis with cirrhosis (Telenti et al., in press). Three months after transplantation, she developed a primary EBV infection (Figure). At 6 months post-transplant, liver function tests were markedly elevated. Histologic evaluation of a liver biopsy was compatible with organ rejection; she received methylprednisolone to control this condition. At 19 months post-transplant, liver function tests were again elevated and histologic examination showed dense portal and intralobular infiltration with mononuclear cells displaying immunoblastic features. She received methylprednisolone and a 10-day course of intravenous anti-T cell monoclonal antibody (OKT3). Liver function tests indicated a deterioration of organ function. At 22 months post-transplant, the diagnosis of EBV hepatitis was considered. Azathioprine was discontinued and acyclovir was administered. Over the following 3 months, infiltrates within the liver began to clear and liver function tests improved.

EBV sequences were amply present in biopsy tissues obtained during the epidsode of acute hepatitis that followed EBV seroconversion (biopsy 2) and at the time of therapy with OKT3 (biopsy 3). EBV was not detected in biopsy 1 (before EBV infection) (Figure). EBV amplification decreased substantially after antiviral chemotherapy. The patient currently is doing well and has only mild abnormalities of liver function tests. This case clearly demonstrates the potential of PCR analysis for monitoring EBV infections in organ transplant patients and the need for the differential diagnosis of infection versus immunologic organ rejection in these patients.

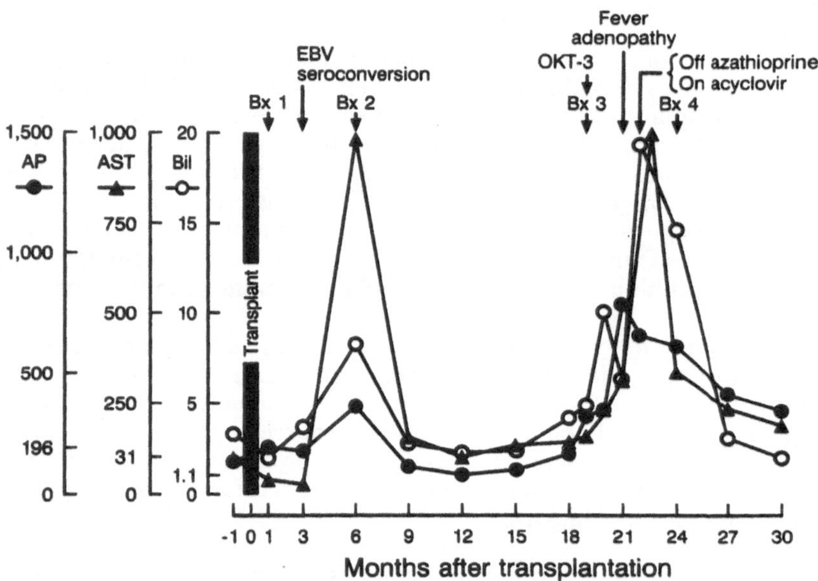

Fig. 1 Laboratory profile of a patient with persistent EBV-related liver graft dysfunction. The time relationship between distinct clinical events and liver abnormalities is marked by arrows. AP = alkaline phosphatase (normal range for females, 76-196 U/L; males, 90-239 U/L; AST = asparats aminotransferase (normal, 12-31 U/L); Bil = total bilirubin (normal , 1.1 mg/ml). Permission for use of figure granted by Hepatology.

SUMMARY

In the past five years, technologic advances in the shell vial assay and expanding availability of rapid membrane EIA tests have allowed over 90% of the viruses detected in our laboratory to be reported within 24 h postinoculation. PCR technology promises to add a new practical dimension to diagnostic virology especially for the detection of viruses in CSF, tissues, and blood. Extension of these diagnostic capabilities from investigative protocols to general laboratories for routine use will be our biggest chellenge and be based on considerations of cost, licensing, and availability of this technology in "kit" formats.

REFERENCES

Agha, S.A., Coleman, J.C., Selwyn, S., Mahmoud, L.A., and Abd-Eloal, A.M., 1988, Combined use of sonication and monoclonal antibodies for the detection of early and late cytomegalovirus antigens in centrifugation culture, J Virol Methods, 22:41.

Aurelius, E., Johansson, B., Sköldenberg, Staland, A. and Forsgren, M. 1991, Rapid diagnosis of herpes simplex encephelitis by nested polymerase chain assay of cerebrospinal fluid, Lancet, 1:89.

Baker, D. A., Gonik, B., Milch, P.O., Berkowitz, A. Lipson, S., and Verma, U., 1989, Clinical evaluation of a new herpes simplex virus ELISA: a rapid diagnositc test for herpes simplex virus, Obste Gynecol., 73:322.

Dascal, A., Chan-Thim, J., Morahan, M., Portnoy, J., and Mendelson, J., 1989, Diagnosis of herpes simplex virus infection in a clinical setting by a direct antigen detection enzyme immunoassay kit, J Clin Microbiol., 27:700.

Dascal, A., Chan-Thim, J., Morahan, M., Portnoy, J., and Mendelson, J., 1989, Replacement of special enzyme immunoassay transport medium by a standard viral transport medium in the Herpcheck herpes simplex virus antigen detection kit, Diagn Microbiol Infect Dis., 12:473.

Dorian, K. J., Beatty, E., and Atterbury, K.E., 1990, Detection of herpes simplex virus by the Kodak SureCell herpes test, J Clin Microbiol., 28:2117.

Dunkel, E.C., Pavan-Langston, D., Fitzpatrick, K., and Cukor, G., 1988, Rapid detection of herpes simplex virus (HSV) antigen in human ocular infections, Curr Eye Res., 7:661.

Espy, M.J., Wold, A.D., Ilstrup, D.M. and Smith, T.F., 1988, Effect of treatment of shell vial cell cultures with dimethyl sulfoxide and dexamethosaone for detection of cytomegalovirus, J Clin Microbiol., 26:1091.

Fedorko, D.P., Ilstrup, D.M., and Smith, T.F., 1988, Effect of age of shell vial monolayers on detection of cytomegalovirus from urine specimens, J Clin Microbiol., 27:2107.

Gleaves, C.A., Smith, T.F., Shuster, E.A., and Pearson, G.R. 1984, Rapid detection of cytomegalovirus in MRC-5 cells inoculated with urine specimen by using low-speed centrifugation and monoclonal antibody to an early antigen, J Clin Microbiol., 19:917.

Gleaves, C.A., Wilson, D.J., Wold, A.D., and Smith, T.F., 1985, Detection and serotyping of herpes simplex virus in MRC-5 cells by use of centrifugation and monoclonal antibodies 16 h postinoculation, J Clin Microbiol., 21:29.

Gray, L.D., Ilstrup, D.M., and Smith, T.F., 1989, Correlation of urine pH with the detection of cytomegalovirus by the shell vial technique, Diagn Microbiol Infect Dis, 12:275.

Hsia, K., Spector, D.H., Lawrie, J., and Spector, S.A., 1989, Enzymatic amplification of human cytomegalovirus sequences by polymerase chain reaction, J Clin Microbiol, 27:1802.

Hudson, J.B., 1988, Further studies on the mechanisms of centrifugal enhancement of cytomegalovirus infectivity, J Virol Methods, 19:97.

Jespersen, D.J., Drew, W.L., Gleaves, C.A., Meyers, J.D., Warford, A.L., and Smith, T.F., 1989, Multisite evaluation of a monoclonal antibody reagent (Syva) for rapid diagnosis of cytomegalovirus in the shell vial assay, J Clin Microbiol., 27:1502.

Jiwa, N.M., Van Gemert, G.W., Raap, A.K. Van de Rijke, F.M., Mulder, A., Lens, P.F., Salimana, M.M., Zwaan, F.E., Van Dorp, W., and Van der Ploeg, M., 1989, Rapid detection of human cytomegalovirus DNA in peripheral blood leukocytes of viremic transplant recipients by the polymerase chain reaction, Transplantation, 48:72.

Langston, D.P., and Dunkel, E.C., 1989, A rapid clinical diagnostic test for herpes simplex infectious hepatitis, Am J Ophthalmol., 107:675.

Lucas, G.J., Seigneurin, J.M., Tamalet, J., Michelson, S., Baccard, M., Delagneau, J.F., and Deletoille, P., 1989, Rapid diagnosis of cytomegalovirus by indirect immunofluorescence assay with monoclonal antibody F6b in a commercially available kit, J Clin Microbiol., 27:367.

McClintock, J.T., Thaker, S.R., Mosher, M., Jones, D., Forman, M., Charache, P., Wright, K., Keiser, J., and Taub, F.E., 1989, Comparison of in situ hybridization and monoclonal antibodies for early detection of cytomegalovirus in cell culture, J Clin Microbiol., 27:1554.

Michalski, F.J., Shaikh, M., Sahraie, F., Desai, S., Verano, L., and Vallanhaneni, J., 1986, Enzyme-linked immunosorbent assay spin amplification technique for herpes simplex virus antigen detection, J Clin Microbiol., 24:310.

Minnich, L.L., Smith, T.F. Ray, C.G., and Specter, S., 1988, Rapid detection of viruses by immunofluorescence, Cumitech, 24:1.

Paya, C.V., Holley, K.E., Wisner, R.H., Balasubramanian, K., Smith, T.F., Espy, M.J., Ludwig, J., Batts, K.P., Hermans, P.E., and Krom, R.A.F. 1990, Early diagnosis of cytomegalovirus hepatitis in liver transplant recipients: role of immunostaining, DNA hybridization and culture of hepatic tissue, Hepatology, 12:119.

Peterson, E.M. Hughes, B.L. Aarnaes, S.L., and de la Maza, L.M., 1988, Comparison of primary rabbit kidney and MRC-5 cells and two stain procedures for herpes simplex virus detection by a shell vial centrifugation method. J Clin Microbiol. 26:222.

Rowley, A.H., R.J., Lakeman, F.D., and Wolinsky, S.M., 1990, Rapid detection of herpes-simplex virus DNA in cerebrospinal fluid of patients with herpes simplex encephalitis, Lancet, 1:440.

Shuster, E.A., Beneke, J.S., Tegtmeier, G.E., Pearson, G.R., Gleaves, C.A., Wold, A.D., and Smith, T.F., 1985, Monoclonal antibody for rapid laboratory detection of cytomegalovirus infections: characterization and diagnostic application, Mayo Clin Proc., 60:577.

Sorbello, A.F., Elmendorf, S.L., McSharry, J.J., Venezia, R.A., and Echols, R.M., 1988, Rapid detection of cytomegalovirus by fluorescent monoclonal antibody staining and in situ DNA hybridization in a dram vial cell culture system, J Clin Microbiol., 26:1111.

Stöckl, E., Popow-Kraupp, T., Heing, F.X., Muhlbacher, F., Balche, P., and Kunz, C., 1988, Potential of in situ hybridization for early diagnosis of productive cytomegalovirus infection, J Clin Microbiol., 26:2536.

Swierkosz, E.M., Flanders, R., Melvin, L., Miller, J.D., and Kline, M.W., 1989, Evaluation of the Abbott TESTPACK RSV enzyme immunoassay for detection of respiratory syncytial virus in nasopharngeal swab specimens, J Clin Microbiol., 27:199.

Tac, P., Aarnes, S.L., de la Maza, L.M., and Peterson, E.M., 1989, Detection of herpes simplex virus by 8 h in shell vial cultures with primary rabbit kidney cell, J Clin Microbiol., 27:199.

Telenti, A., Marshall, W.F., and Smith, T.F., 1990, Detection of Epstein-Barr virus by polymerase chain reaction, J Clin Microbiol., 28:2187.

Telenti, A., Smith, T.F., Ludwig, J., Keating, M.R., Krom, R.A.F., and Wiesner, R.H., 1991, Epstein-Barr virus and persistent graft dysfunction following liver transplantation. Hepatology, 14:282.

Thiele, G.M., and Woods, G.L., 1988, The effect of dexamethasone on the detection of cytomegalovirus in tissue culture and by immunofluorescence, J Virol Methods, 22:319.

Van der Bij, W., Schirm, J., Torensma, R., van Son, W.J., Tegzess, A.M., and Hauh-The, T., 1988, Comparison between viremia and antigenemia for detection of cytomegalovirus in blood, J Clin Microbiol., 26:2531.

Van der Bij, W., van Son, W.J., van der Berg, A.P.M., Tegzess, A.M., Torensma, R. and The, T.H., 1989, Cytomegalovirus (CMV) antigenemia: rapid diagnosis and relationship with CMV-associated clinical syndromes in renal allograft recipients, Transplant Proc., 21:2061.

Waner, J.L., Todd, S.J., Shalaby, H., Murphy, P., and Wall, L.V., 1991, Comparison of Directigen FLU-A with viral isolation and direct immunofluorescence for the rapid detection and identification of influenza A virus, J Clin Microbiol., 29:479.

West, P.G., Aldrich, B., Hartwig, R., and Haller, G.J, 1988, Enhanced detection of cytomegalovirus in confluent MRC-5 cells treated with dexamethasone and dimethyl sulfoxide, J Clin Microbiol., 26:2510.

Wu, T.C., Zaza, S., and Callaway, J., 1989, Evaluation of the DuPont HERPCHEK herpes simplex virus antigen test with clinical specimens, J Clin Microbiol., 27:1903.

Wunderli, W., Kagi, Gruter, and Auracher, J.D., 1989, Detection of cytomegalovirus in peripheral leukocytes by different methods, J Clin Microbiol., 27:1916.

Ziegler, T., Waris, M., Rautiainen, M., and Arstila, P., 1988, Herpes simplex virus detection by macroscopic reading after overnight incubation and immunoperoxidase staining, J Clin Microbiol., 26:2013.

DNA PROBES FOR VIRAL DIAGNOSIS

Harley A. Rotbart

University of Colorado
School of Medicine
Denver, CO 80262

SUMMARY AND INTRODUCTION

Great excitement has surrounded the development of molecular hybridization techniques for viruses because of the potential of nucleic acid probes as diagnostic reagents. The need for rapid and accurate viral diagnosis has become particularly pressing with the recent availability of numerous drugs which have been clinically proven as effective in antiviral chemotherapy and prophylaxis. Additionally, even for those viruses for which no therapy is currently available or for which none is needed, the ability to diagnose a causative viral pathogen in a patient with a disease of protean manifestations and multiple possible etiologies is of value. Many such patients are routinely treated empirically with unnecessary antibacterial or antifungal drugs until bacterial and/or fungal disease can be ruled out. Proof of viral etiology in such patients will reduce unnecessary therapy and even unnecessary hospitalization. Furthermore, identification of a specific viral pathogen will often allow the physician to predict the patient's prognosis and need for future care. Viral diagnosis also plays an important role in epidemiology and hospital infection control, where knowledge of the specific infecting virus is critical to characterizing and containing the spread of the virus from person to person. Finally, viral diagnosis can be both money saving and money earning for a hospital and clinic.

There are numerous circumstances in which nucleic acid probes should be advantageous in the diagnosis of viral infections. Many viruses are difficult or impossible to grow in standard tissue culture or, if they do grow, may take many days to weeks before presenting the typical cytopathic effect required for detection. Often members of a single group of viruses lack a common antigen which can be exploited for immunoassay techniques; in such instances, dozens of separate assays might be required to diagnose or rule out that group of agents (e.g. enteroviruses, rhinoviruses). Serologic studies are most useful retrospectively (by comparing acute and convalescent titers of IgG antibody) and only when the specific serotype of infecting virus is known or suspected; IgM and IgA assays can be problematic and are frequently unreliable. Probes for nucleic acid are the most accurate way of intracellular localization of viruses (via *in situ* hybridization). Finally, individual laboratory circumstances may make probes more practical than tissue culture, particularly when kit formats incorporating hybridization technology become widely available; use of hybridization

Innovations in Antiviral Development and the Detection of Virus Infection
Edited by T. Block *et al.*, Plenum Press, New York, 1992

for accelerating detection of tissue culture results or confirming other assay results are additional considerations for the individual laboratory.

PRINCIPLES OF NUCLEIC ACID HYBRIDIZATION

The basic concepts of hybridization have been known for many years. Briefly summarized:

1. Complementary single strands of DNA or RNA will anneal with each other according to second order kinetics.

2. The rate of annealing depends upon:
   a) length of the "probe" molecule
   b) number of complementary sequences
   c) temperature of the reaction
   d) ionic strength (minimal effect at >0.4 M)
   e) viscosity (e.g. dextran sulfate, polyethylene glycol)
   f) pH (this has only a minimal effect between pH 5-9 with adequate salt)

3. The stability of the hybrids which are formed depends upon the "stringency" of hybridization conditions. Stringency is determined by:
   a) temperature
   b) base pair composition
   c) ionic strength of buffer
   d) concentration of denaturants (e.g. formamide)

   The "melting" or denaturation temperature of hybrids is the T_m, and

   $$T_m = 81.5° + 16.6 \log M + 0.41 (\%G+C) - 500/n - 0.61 (\%formamide)$$

   where M = ionic strength and (G+C) represents the relative content of those two bases.

4. The ideal temperature (maximum rate and stability) for hybridization of large probes (>150 nucleotides) is:

   $$T_m - 25°C \text{ (at 1.0 M salt)}$$

   where T_m = denaturation temperature for the hybrid

5. The ideal temperature for hybridization of small probes is:

   $$T_m - 5°C \text{ (if perfectly matched)}$$

   where $T_m = 4(G+C) + 2(A+T)$ at 1 M salt

6. For hybrids >150 nucleotides, the T_m decreases by 1°C/1% mismatch

7. For hybrids <20 nucleotides, the T_m decreases by 5°C/mismatch

8. Maximum stringency = maximum specificity; conversely, minimum stringency = maximum sensitivity.

APPLICATIONS OF DNA PROBES IN VIRAL DIAGNOSIS

In the research laboratory setting, great successes have been reported for nucleic acid hybridization-based viral detection. Simply said, any and all viruses which can be partially cloned, or for which partial sequence information is known, can be successfully detected by hybridization in the laboratory; hundreds of such reports have appeared. The viruses which are detected are usually tissue culture preparations, occasionally spiked back into sterile body fluids in reconstruction type experiments. The usual next step is to attempt clinical diagnosis using techniques which worked well in the experimental setting. It is at that juncture that the great disappointments in nucleic acid hybridization technology have occurred. To date, virtually no successful hybridization assays for viruses have been marketed for direct (i.e. without first "passing" the virus in tissue culture) testing of clinical specimens. The reasons for the many failures at making the transition from the laboratory to the clinical arena are threefold: inadequate sensitivity, inadequate specificity, and inconvenience.

Sensitivity

The two factors which can adversely affect the sensitivity of a hybridization assay are sub-optimal specimens and/or sub-optimal probes. In the case of viruses which are predominantly cell associated, certain specimens may have too few cells to be of value. Where virus is primarily cell-free, the inoculum in a particular body fluid may be too low. Finally, many body fluids and tissues contain substances which can inhibit the hybridization reaction; these include nucleases which can break down both target and probe, as well as proteins and lipids which can interfere with the reaction kinetics. The failure of certain probes, on the other hand, may simply be the result of weak homology with the target molecules. Additionally, the length of the probe may be an important factor - oligomeric probes, for example, often provide as little as 1/100 the sensitivity of longer probes for the same target. The detector system is a further limiting variable in the sensitivity of hybridization probes, with current color-labelling techniques shown to be 1/10 - 1/100 as sensitive as those using ^{32}P. In most reported systems, the limit of sensitivity of nucleic acid hybridization (using ^{32}P-labeled probes) for viral detection is approximately 1 pg of purified viral RNA or DNA. That translates into between 10^4 and 10^6 virions. In cerebrospinal fluid, however, as few as 10-1000 virions per ml may be the rule in, for example, enteroviral meningitis. Similarly low inocula of many viruses may be found in serum and urine, making direct testing of body fluids difficult.

Specificity

There are several routes to false positivity in hybridization reactions. Firstly, certain viruses, e.g. herpesviruses and enteroviruses, are known to share genomic sequences with mammalian DNA and RNA; hence, the unwitting selection of viral probes from regions which cross-react with host nucleic acid can result in true hybridization, but a false positive assay for the viruses in question. Homology between vector sequences and host nucleic acid may also result in poor specificity, a problem unique to systems in which the vector is labeled along with the probe insert and included in the hybridization reaction (as is the standard procedure for nick-translated cDNA probes). G-C rich regions of host or contaminating DNA may provide yet another source of false-positives, with non-specific binding of probes which have similar G-C content. Finally, contaminating proteins and lipids, as well as the filter paper on which the target nucleic acid is applied, may all provide non-specific and sticky landing sites for labeled probe sequences.

Convenience

Currently-designed hybridization systems can be quite cumbersome. Certainly assays requiring radioisotopes are not user-friendly. In addition, testing of clinical specimens generally requires an extraction step to "clean up" the sample; this procedure removes contaminating substances such as nucleases, other proteins, and lipids. Additionally, extraction procedures inactivate live viruses and standardize specimen handling. These desirable features are offset by the extra processing time required as well as the cumbersome handling of hazardous chemicals used in extracting nucleic acids.

THE "HYBRIDIZATION DECISION TREE"

Unfortunately, the variables of sensitivity, specificity, and convenience each operate in conflict with the others, as shown in this schematic:

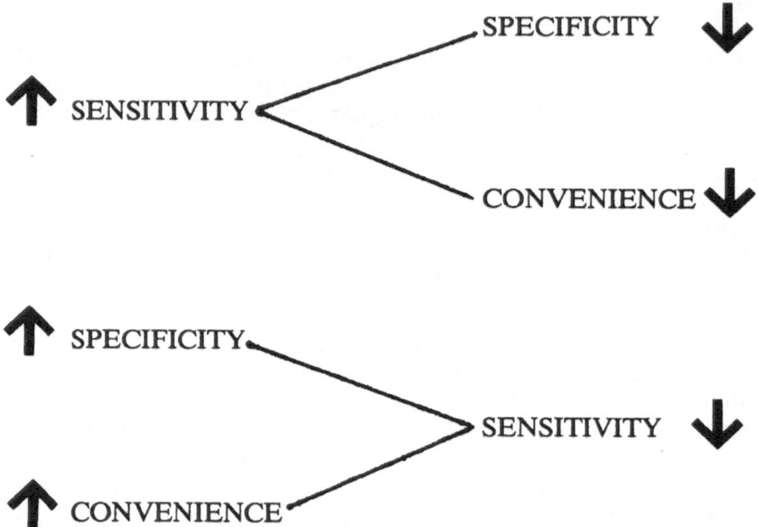

In designing a hybridization assay for viruses, it is therefore useful to consider each of the parameters in the assay separately for their relative merits of sensitivity, specificity, and convenience. The following "hybridization decision tree" summarizes these parameters:

HYBRIDIZATION DECISION TREE

PROBE

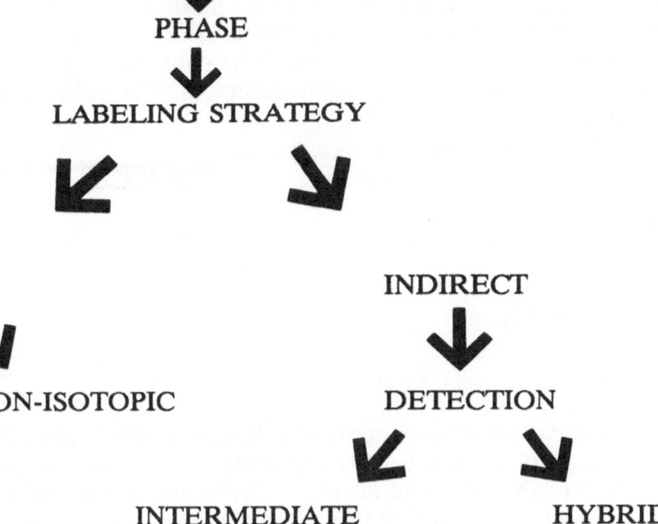

PHASE

LABELING STRATEGY

DIRECT

INDIRECT

RADIOACTIVE

NON-ISOTOPIC

DETECTION

INTERMEDIATE

HYBRID

DETECTOR/CONJUGATE

SUBSTRATE
RESPONSE

The first variable on the "tree" to be considered is the type of probe to be used. Several options exist, including complementary DNA (cDNA) probes labeled by nick-translation single-stranded RNA probes labeled during *in vitro* transcription, and oligomeric probes synthesized to specifications and end-labelled with a standard kinase reaction. The relative merits of each of these types of probes are shown below, where the more "+"'s, the better a probe type is in that particular parameter.

TYPE	SENSITIVITY	SPECIFICITY	CONVENIENCE
cDNA	+ +	+ +	+
RNA	+ + +	+	+ +
oligo	+	+ + +	+ + +

Thus, RNA probes are most sensitive while oligomeric probes are the most specific and convenient. Unfortunately, RNA probes are most prone to giving false-positive reactions and oligomeric probes most likely to yield false-negative results. cDNA probes, while being the least convenient (most cumbersome) to prepare and use, have intermediate sensitivity and specificity compared with RNA and oligomeric probe counterparts.

Several alternatives exist for the "phase" in which the hybridization reaction is performed. Two strands of nucleic acid will hybridize to each other when both are in solution (liquid phase) or when one is in solution and the other affixed to a support such as filter paper (solid phase). "Sandwich" hybridizations represent a cross between liquid and solid phases (mixed phase), where a first probe reacts with the target in solution and a second probe (to a different or proximal portion of the target sequences), affixed to a filter paper or column support "captures" the hybrid onto the support where it is subsequently detected. The relative merits of these phase options are shown below:

TYPE	SENSITIVITY	SPECIFICITY	CONVENIENCE
SOLID	+++	++	+++
MIXED (SANDWICH)	+	+++	++
LIQUID	+++	+++	+

The next important decision in designing a hybridization assay is the probe labeling strategy. Probes may be directly labeled, meaning the marker molecule is chemically attached to the probe molecule. Alternatively, an indirect labeling approach may be taken, in which the probe itself is not labeled with a detectable entity, but a secondary reaction is performed to detect the hybrid. Two general options exist for directly labeled probes, radioactive and non-radioactive, with their relative merits shown below:

TYPE	SENSITIVITY	SPECIFICITY	CONVENIENCE
Direct Labels			
Radioactive	+++	+++	+
Non-isotopic	+	++	+++

Dozens of indirect labeling schemes for DNA probes have been reported in recent years, making straightforward comparisons impossible. Some of the available options are shown below; they are categorized based on whether the detection is of the hybrid itself or of an intermediate compound bound to the probe. While biotin is the most frequently used intermediate, many others have been reported.

INDIRECT LABELS

Hybrid detection

Anti-DNA/DNA antibody
Anti-DNA/RNA antibody
Histone

Intermediate compound detection

biotin		ethidium
sulfone		DNP
poly dT	PROBE	glucosyl
poly dA		Lac operon DNA
S peptide		mercuric cyanide

Finally, with either directly labeled non-isotopic probes or indirectly labeled probes (all of which are non-isotopic, by definition), a detection system is employed in which a substrate is converted into either color or light-producing substances. This substrate response also presents a decision of relative merits.

TYPE	SENSITIVITY	SPECIFICITY	CONVENIENCE
COLOR	+	(+)	+++
LIGHT	+++	(++)	+

The parentheses indicate that comparisons of the specificity of these two substrate response options is imprecise because of limited applications of these methods to date.

BRINGING VIRAL DNA PROBES TO MARKET

Two important marketing considerations are prerequisites for private enterprise to commit to producing hybridization kits for viral diagnosis: 1) will such kits be cost effective? and 2) will hybridization outperform immunoassays for those pathogens for which the latter are currently or imminently available? Dr. Fred Tenover addressed the first of those issues in a recent review (see references). His analysis shows that a hybridization assay could be performed at a cost of less than $10.00 per specimen if more than five specimens were run per day; with fewer number of specimens to be tested, the cost per specimen might be prohibitive. The specifics of those calculations follow:

Cost of kit (20 filters)	$ 320.00
shelf life of kit = 30 days	
Additional reagents (20 filters)	$ 100.00
Material cost per filter ($420/20)	$ 21.00
Labor cost per filter (@ $13/hour)	$ 19.50
(One-time equipment expense)	$ 3,500.00

Volume of specimens tested	Cost per specimen
1 specimen/day	$40.50
5 specimens/day	$ 8.10
50 specimens/day	$ .81

A number of studies have directly compared hybridization based viral assays with commercially available or research stage immunoassays for the same pathogens. These comparisons, including herpes simplex virus, parvovirus, rotavirus, and respiratory syncytial virus, have all concluded that immunoassays are at least as sensitive and specific, and often times more sensitive and specific, than parallel hybridization assays. Until improvements can be made in our approach to probe-based testing, its usefulness may be limited to those pathogens for which immunoassays cannot be successfully developed.

Much as immunoassays have an intrinsic limit of achievable sensitivity, it appears that hybridization-based assays may also have reached their "brick wall." Everything discussed in this review, of course, refers to standard approaches to nucleic acid hybridization. The chapter which follows this one, however, discusses the new technological breakthrough which promises to resuscitate the hybridization type of approach--the amplification of target viral genome to a level readily detectable by standard hybridization assays without needing to reduce stringency and adversely impact specificity. The polymerase chain reaction and other amplification strategies promise to make nucleic acid detection the pre-eminent viral diagnostic modality of the coming decade.

In conclusion,

1. Nucleic acid hybridization holds great theoretical promise in the field of diagnostic virology.

2 Research applications have far outdistanced clinical applications to date.

3. Major limitations to adaptation of probes to the clinical arena include inadequate sensitivity and specificity, and technical inconvenience.

4. Numerous advances have been made in the design of newer probes and detection systems which begin to address the shortcomings of hybridization.

5. Sensitivity directly affects both specificity and technical convenience. Independent resolution of the sensitivity limitations (e.g. by target amplification) will reduce the problems of specificity and convenience.

REFERENCES

Ambinder, R.F., Charche, P., Staal, S., Wright, P., Forman, M., Hayward, S.D., and Hayward, G.S., 1986, The vector homology problem in diagnostic nucleic acid hybridization of clinical specimens, J. Clin. Microbiol. 24:16-20.

Chernesky, M.A., 1989, An evolutionary change in diagnostic virology, Yale J. Biol. Med. 62:89-92.

Edberg, S.C., 1985, Principles of nucleic acid hybridization and comparison with monoclonal antibody technology for the diagnosis of infectious diseases, Yale J. Biol. Med. 58:425-442.

Eisenstein, B.I., 1990, New molecular techniques for microbial epidemiology and the diagnosis of infectious diseases, J. Infect. Dis. 161:595-602.

Geiger, R., Hauber, R., and Miska, W., 1989, New bioluminescence-enhanced detection systems for use in enzyme activity tests, enzyme immunoassays, protein blotting and nucleic acid hybridization, Mol. Cell Probes 3:309-328.

Hsiung, G.D., 1989, The impact of cell culture sensitivity on rapid viral diagnosis: A historical perspective, Yale J. Biol. Med. 62:79-88.

Landry, M.L., Mayo, D.R., and Hsiung, G.D., 1989, Rapid and accurate viral diagnosis, Pharmac. Ther. 2:287-328.

Matthews, J.A., Batki, A., Hynds, C., and Kricka, L.J., 1985, Enhanced chemiluminescent method for the detection of DNA dot-hybridization assays, Anal. Biochem. 151:205-209.

Matthews, J.A., and Kricka, L.J., 1988, Analytical strategies for the use of DNA probes, Anal. Biochem. 169:1-25.

Meinkoth, J., and Wahl, G., 1984, Hybridization of nucleic acids immobilized on solid supports, Anal. Biochem. 138:267-284.

Norval, M., and Bingham, R.W., 1987, Advances in the use of nucleic acid probes in diagnosis of viral diseases of man, Arch. Virol. 97:151-165.

Richman, D.D., 1987, Developments in rapid viral diagnosis, Infect. Dis. Clin. NA 1:311-322.
Richman, D.D., Cleveland, P.H., Redfield, D.C., Oxman, M.N., and Wahl, G.M., 1984, Rapid viral diagnosis, J. Infect. Dis. 149:298-310.
Siegler, N., 1989, DNA-based testing: A progress report, ASM News 55:308-312.
Tenover, F.C., 1988, Diagnostic deoxyribonucleic acid probes for infectious diseases, Clin. Microbiol. Rev. 1:82-101.
Thomas, P.S., 1980, Hybridization of denatured RNA and small DNA fragments transferred to nitrocellulose, Proc. Natl. Acad. Sci. USA 77:5201-5205.
Van Brunt, J., and Klausner, A., 1987, Pushing probes to market, Biotechnology 5:211-221.
Viscidi, R.P., and Yolken, R.G., 1987, Molecular diagnosis of infectious diseases by nucleic acid hybridization, Mol. Cell Probes 1:3-14.
Wahl, G.M., Stern, M., and Stark, G.R., 1979, Efficient transfer of large DNA fragments from agarose gels to diazobenzyloxymethyl-paper and rapid hybridization by using dextran sulfate, Proc. Natl. Acad. Sci. USA 76:3683-3687.
White, B.A., and Bancroft, F.C., 1982, Cytoplasmic dot hybridization: Simple analysis of relative mRNA levels in multiple small cell or tissue samples, J. Biol. Chem. 257:8569-8572.
Yolken, R.H., 1988, Nucleic acids or immunoglobulins: which are the molecular probes of the future?, Mol. Cell Probes 2:87-96.
Yolken, R.H., Coutlee, F., and Viscidi, R.P., 1989, New prospects for the diagnosis of viral infections, Yale J. Biol. Med. 62:131-139.

CONTRIBUTORS

ASHORN, PER
Laboratory of Viral Diseases
National Institute of Allergy
and Infectious Diseases
National Institute of Health
Bethesda, Maryland

BARIGHT, DONALD E.
Department of Virology
Sterling Research Group
Rensselaer, New York

BERGER, EDWARD A.
Laboratory of Viral Diseases
National Institute of Allergy
and Infectious Diseases
National Institute of Health
Bethesda, Maryland

BLOCK, TIMOTHY M.
Department of Microbiology
and Immunology
Jefferson Medical College
Thomas Jefferson University
Philadelphia, Pennsylvania

BRIGANDI, RICHARD A.
Department of Microbiology
and Immunology
Jefferson Medical College
Thomas Jefferson University
Philadelphia, Pennsylvania

CANTIN, EDOUARD M.
City of Hope National Medical Center
Department of Neurology
Duarte, California

CHAMBERS, LORENA
Department of Pharmacy
Thomas Jefferson University
Philadelphia, Pennsylvania

CHENG, Y.S. EDMOND
DuPont Merck Pharmaceutical Company
Wilmington, Delaware

COEN, DONALD M.
Department of Biological Chemistry
and Molecular Pharmacology
Harvard Medical School
Boston, Massachusetts

COLONNO, RICHARD J.
Department of Virus
and Cell Biology
Merck Sharp & Dohme
Research Laboratories
West Point, Pennsylvania

COTE, PAUL J.
Division of Molecular Virology
and Immunology
Department of Microbiology
Georgetown University
Rockville, Maryland

DE BOECK, HILDE
Bristol Myers-Squibb
Pharmaceutical Research Institute
Wallingford, Connecticut

DIANA, GUY D.
Department of Medicinal Chemistry
Sterling Research Group
Rensselaer, New York

DI NELLO, ROBERT
Chiron Corporation
Emeryville, California

DIRENZO, SUE
Department of Pathology
Thomas Jefferson University
Philadelphia, Pennsylvania

DURANTE, MARYANN
Department of Pathology
Thomas Jefferson University
Philadelphia, Pennsylvania

DUTKO, FRANK J.
Department of Virology
Sterling Research Group
Rensselaer, New York

EISENSTEIN, TOBY K.
Department of Microbiology
and Immunology
Temple University School of Medicine
Philadelphia, Pennsylvania

EPSY, MARK J.
Section of Clinical Microbiology
Mayo Clinic
Rochester, Minnesota

FIELDS, BERNARD N.
Harvard Medical School
Boston, Massachusetts

FOX, M. PAT
Department of Virology
Sterling Research Group
Rensselaer, New York

FRANCIS, BRIAN
Ortho Diagnostic Systems, Inc.
Raritan, New Jersey

FRIEDMAN, LAWRENCE S.
Division of Gastroenterology
and Hepatology
Jefferson Medical College
Philadelphia, Pennsylvania

GERIN, JOHN L.
Division of Molecular Virology
and Immunology
Department of Microbiology
Georgetown University
Rockville, Maryland

GRAFSTROM, ROBERT H.
Department of Microbiology
and Immunology
Jefferson Medical College
Thomas Jefferson University
Philadelphia, Pennsylvania

HICKS, KAREN A.
Department of Biological Chemistry
and Molecular Pharmacology
Harvard Medical School
Boston, Massachusetts

HITCHCOCK, MICHAEL J.M.
Bristol Myers-Squibb
Pharmaceutical Research Institute
Wallingford, Connecticut

HO, HSU-TSO
Bristol Myers-Squibb
Pharmaceutical Research Institute
Wallingford, Connecticut

HOPKINS, LEIGH
Department of Pharmacy
Thomas Jefferson University
Philadelphia, Pennsylvania

HSIEH, SEN-YUNG
Fox Chase Cancer Center
Philadelphia, Pennsylvania

INGEMARSON, ROLF
Department of Biochemistry
Bio-Mega Inc.
2100 rue Cunrad
Laval, Quebec, Canada

JUNGKIND, DONALD
Department of Pathology
Thomas Jefferson University
Philadelphia, Pennsylvania

KONRAD, SHERRY A.
Bristol Myers-Squibb
Pharmaceutical Research Institute
Wallingford, Connecticut

KORANT, BRUCE D.
DuPont Merck Pharmaceutical Company
Wilmington, Delaware

KORBA, BRENT E.
Division of Molecular Virology
and Immunology
Department of Microbiology
Georgetown University
Rockville, Maryland

LEE, STEPHEN
Ortho Diagnostic Systems, Inc.
Raritan, New Jersey

LIUZZI, MICHEL
Department of Biochemistry
Bio-Mega, Inc.
2100 rue Cunrad
Laval, Quebec, Canada

MARTIN, PAUL
Division of Gastroenterology
and Hepatology
Jefferson Medical College
Philadelphia, Pennsylvania

MC HUTCHINSON, JOHN
Division of GI and Liver Disease
University of Southern California
Los Angeles, California

MC KINLAY, MARK A.
Department of Virology
Sterling Research Group
Rensselaer, New York

MERLUZZI, VINCENT J.
Boehringer Ingelheim
Pharmaceuticals, Inc.
Ridgefield, Connecticut

MORAHAN, PAGE S.
Department of Microbiology
and Immunology
The Medical College of Pennsylvania
Philadelphia, Pennsylvania

MOSS, BERNARD
Laboratory of Viral Diseases
National Institute of Allergy
and Infectious Diseases
National Institute of Health
Bethesda, Maryland

NELLES, Mitchell
Ortho Diagnostic Systems, Inc.
Raritan, New Jersey

OPENSHAW HARRY
City of Hope National Medical Center
Department of Neurology
Duarte, California

OTTO, MICHAEL J.
DuPont Merck Pharmaceutical Co.
Wilmington, Delaware

PATTERSON, CATHERINE E.
DuPont Merck Pharmaceutical Co.
Wilmington, Delaware

PELOSI, EMANUELA
Department of Biological Chemistry
and Molecular Pharmacology
Harvard Medical School
Boston, Massachusetts

PERVEAR, DANIEL C.
Department of Virology
Sterling Research Group
Rensselaer, New York

PODSAKOFF, GREGORY
City of Hope Medical Center
Department of Neurology
Duarte, California

POLITO, ALAN
Chiron Corporation
Emeryville, California

QUAN, STELLA
Chiron Corporation
Emeryville, California

RINEHART, KENNETH L.
Department of chemistry
University of Illinois
Urbana, Illinois

RIZZO, CHRISTOPHER J.
DuPont Merck Pharmaceutical Co.
Wilmington, Delaware

ROSENTHAL, ALAN S.
Boehringer Ingelheim
Pharmaceuticals, Inc.
Ridgefield, Connecticut

ROSSI, JOHN J.
Department of Molecular Genetics
Beckman Research Institute
of the City of Hope
Duarte, California

ROTBART, HARLEY A.
University of Colorado
School of Medicine
Denver, Colorado

RUCKER, RONALD G.
DuPont Merck Pharmaceutical Co.
Wilmington, Delaware

SACKS, STEPHEN L.
Division of Infectious Diseases
University of British Columbia
Vancouver, British Columbia, Canada

SARVER, NAVA
Developmental Therapeutics Branch
Division of AIDS, NIAID
Bethesda, Maryland

SCOUTEN, ERIKA
Department of Biochemistry
Bio-Mega Inc.
2100 rue Cunrad
Laval, Quebec, Canada

SMITH, THOMAS F.
Section of Clinical Microbiology
Mayo Clinic
Rochester, Minnesota

TAYLOR, JOHN
Fox chase Cancer Center
Philadelphia, Pennsylvania

TENNANT, BUD C.
Department of Clinical Sciences
College of Veterinary Medicine
Cornell University
Ithaca, New York

WILLEY, DRU E.
City of Hope National Medical Center
Department of Neurology
Duarte, California

WOLD, ARLO D.
Section of Clinical Microbiology
Mayo Clinic
Rochester, Minnesota

WOODS, KATHLEEN L.
Bristol Myers-Squibb
Pharmaceutical Research Institute
Wallingford, Connecticut

WORDELL, CINDY
Department of Pharmacy
Thomas Jefferson University
Philadelphia, Pennsylvania

WU, LINXIAN
Department of Microbiology
and Immunology
The Medical College of Pennsylvania
Philadelphia, Pennsylvania

ZACHARIASEWYCZ, KATHERINE
Department of Microbiology
and Immunology
The Medical College of Pennsylvania
Philadelphia, Pennsylvania

INDEX